神韵
新中式空间设计典藏

Romantic Charm
Collection of New Chinese Space Design

北京天潞诚图书有限公司
广州市唐艺文化传播有限公司　策划

海燕　编

华中科技大学出版社
http://www.hustp.com

contents

目录

第一章 会所空间
The First Chapter Club Space

第二章 住宅空间
The Second Chapter Residential Space

第三章 售楼空间
The Third Chapter Sales Offices Space

第四章 餐饮空间
The Forth Chapter Dining Space

第五章 办公空间
The Fifth Chapter Offices Space

第一章 会所空间
The First Chapter Club Space

012
中华文化盛宴
臻会所
China Cultural Feast
Excellence Club

028
中国红的现代演绎
融汇半岛会所
The Modern Interpretation of China Red
Integrate Peninsula Club

034
茶者，南方之嘉木也
金泰来名车茗茶坊
Tea, also South of the Wood
Kingtailai Cars Tea Club

042
东坡文化新诠释
汤泉会馆
Dongpo Cultural New Interpretation
Tangquan Hall

054
古典中国的印象
北京龙潭湖九五书院
Classical China Impression
Beijing Longtan Lake Paramount Chamber

060
纳西古民的院落生活
束河元年
Naxi Ancient Civilian Courtyard living
Shuhe First Year

064
闲坐庭前细品茗
顺义号茶艺馆
Sit Before the Court of Fine Tea
Shunyi Art Tea House

072
八闽首府之幽幽岁月
聚春园驿馆
The Capital of Fujian in the Faint Years
Meizhou Dongpo Restaurant Yizhuang Shop

082
诗情画意醉东方
福临门会所
Oriental Poetry Drunk
Fook Lam Moon Club

094
阿弥陀佛世界
泰•自然养生会所
Amitabha World
Thai• Natural Health Club

106
婉约徽派，中国画里的乡村
北京寿州大饭店SPA养生会所
Graceful Hui Style, Chinese Painting inside the Country
SPA Health Club of Beijing Shouzhou Grand Hotel

114
巴风蜀韵
成都天府高尔夫球会所
Bashu Charm
Chengdu Tianfu Golf Chamber

120
诗意江南
某茶楼设计
Poetic Jiangnan
A Teahouse Design

124
方圆之道
沁心轩茶会所
The Truth in Square and Circle
Refreshing Hin Tea Club

130
印象茶文化
某茶楼设计
Impression Tea Culture
A Teahouse design

138
水墨画乡
梅林阁会馆
Ink and Wash Paintings
Plum Pavilion

146
琵琶行
水墨江南会所
Song of the Pipa
Ink Jiangnan Club

150
高山流水的韵律之美
三明会所
The Beauty of Rhythm Water
Sanming Club

200
海派禅风
绿地海域观园别墅样板房
Shanghai Zen View
Green Sea View Garden Villa Show Flat

154
东方情韵中的西式优雅
惠州高尔夫会所
Oriental charm of Western Elegance
Huizhou Golf Club

158
思绪的留白
意兰亭休闲会所
Mind Blank
Yi Lanting Leisure Club

164
传统艺术的现代演绎
和谐苑餐饮会所
Modern Interpretation of Traditional Art
Harmony Court Restaurant Club

172
后巷书生，感于斯文
琴南书院
Hou Xiang Scholar, the Feeling in the Gentle
Qin'nan Academy

第二章 住宅空间
The Second Chapter : Residential Space

178
亦古亦今 满室芳华
中星红庐65#别墅
Ancient and Modern Full Chamber Youth
Star Red House 65 # Villa

190
蓝调贵族
融侨新城珑郡别墅样板房
Blues Aristocracy
Rongqiao Legend Villa Show Flat

210
岭南气质
保利越秀•岭南林语F2别墅
Lingnan Temperament
Yuexiu Poly •South of Lingnan Villa F2

216
原木清风
赣州中洋公园首府样板房设计
Log and Breeze
Ganzhou Zhongyang Park Capital Show Flat Design

222
清雅灵韵 内敛悠长
深圳•红树湾现代风格别墅样板房
Elegant Aura,Introverted Long
Shenzhen•Mangrove Bay Modern Style Villa Show Flat

228
静语凝思
上海长城珑湾样板房
Static Language Meditation
Shanghai Great Wall of Long Bay Open Model Houses

234
梅花三弄
央筑花园洋房样板房设计
Plum-blossom in Three Movements
Central Building Garden Villa Flat Design

240
泼墨与白描的新诠释
中航复式A2-1
The New Interpretation of Ink and Line Drawing
CATIC Duplex Apartment A2-1

250
如诗梦，似古今
中央公园二期11-B01
Such as Poem and Dream,Like Ancient and Modern
Central Park, Phase 2, 11-B01

256
沉浸在陶然时光里
正祥橘郡落地样板A户型
Immersed in the Intoxicating time
Zhengxiang Xiangxie Ballet Flat Type A

262
京华烟云
沿海地产赛洛城样板间
A Moment in Beijing
Los Angeles Showroom in the Coastal Real Estate

266
古韵新律
沿海地产赛洛城样板间
Ancient Charm and New Melody
Yangzhu Garden, Chinese Style Show Flat

270
琵琶曲
天悦湾B栋3号样板房
Pipa Tune
Bay B Building No. 3 room Rosedale

276
纯美禅意印象
中洲央筑花园样板房
Pure Beauty Zen Impression
Zhongzhou Central Building Garden Show Flat

284
丹青墨影
宁波交通自在城黄宅
Ink Painting Shadow
Ningbo Traffic Comfortable City Huang Zhai

292
幽梦依稀淡如雪
水月周庄某宅
Like Dream Like Snow
One Residential in Moon Water Zhouzhuang

298
云罗雁飞江南秋
水月周庄某宅
Jiangnan Qiu Yun Luo Yanfe
One Residential in Moon Water Zhouzhuang

304
人境外，听春禽
恒信.春秋府D户型样板房
People Overseas, Listen to the Spring Bird
Hanson• spring and Autumn D Showflat

310
墨语
深房传麒尚林1栋B4样板房
Ink language
Legend Scenery Building 1 B4 Show Flat

316
老洋房的海派风情
上海滩花园洋房
Old villa contains the Shanghai style
Shanghai Beach garden villa

322
山水有禅意
盛世嘉园某宅
Landscape Contains A Zen
A Residence in Shengshi Jiayuan

328
陶瓷智慧与刺绣文化
鹤山十里方圆
Ceramic Wisdom and Embroidery Culture
He Shan "Ten Miles" Villa

334
东方神韵 绝色风华
涛景湾豪宅
Oriental Verve, Beauty and Elegance
Taojingwan Mansion

338
禅意东南亚
嘉宝梦之湾
The Zen of Southeast Asia
Jiabao Dream Bay Show Flat

346
一镜一禅心
泰禾红树林联排别墅
One Mirror, One Zen Heart
Taihe Mangrove Townhouse

350
写意东方 古韵今生
大儒世家卧虎2#305
To Freehand Oriental Rhyme Life
Daru Family Crouching Tiger 2#305

第三章 售楼空间
The Third Chapter : Sales Offices Space

358
密不透风、疏可走马
苍海一墅售楼部
Dense Yet Sparse
The Villa in Aomi Sales Department

364
竹意清幽
江湖禅语销售中心
Bamboo Quiet
Rivers and lakes of Zen Sales Center

374
悠悠古香 淡淡神韵
五洲世纪城售楼处设计
Long Ancient Incense,Subtle Charm
Wuzhou Century City Sales Offices Design

380
东南亚的新东方主义
清澜半岛销售会所
The New Orientalism in Southeast Asia
Qinglan Peninsula Club

390
水墨东钱湖
悦府一期高端私人会所
Ink painting of Dongqian Lake
A High-end Private Club in Yue Fu

第四章 餐饮空间
The Forth Chapter : Dining Space

400
水墨哲学
眉州东坡酒楼亦庄店
Ink Style Philosophy
Meizhou Dongpo Restaurant Yizhuang Shop

406
好客山东
郑州大风餐饮店
Friendly Shandong
Zhengzhou Da Feng Restaurant

410
桥亭回廊间旧梦重温
桥亭活鱼小镇
Renew Romance in Bridge Pavilion Gallery
Bridge Pavilion Fish Town

416
大宋情怀
工三便宜坊
Song Dynasty's feelings
Gongsan Bianyifang

428
浪漫樱花
赤坂日本料理
Romantic Cherry Blossoms
Akasaka Japanese Cuisine

434
春天来了
春天里新川式健康火锅
Spring is Coming
Chuntianli New Sichuan Health Hot Pot

440
锦绣中国红
福州海通一号（梅峰店）
Fairview Chinese Red
Fuzhou Haitong One (Meifeng Shop)

446
黄河谣
中华国宴
The Yellow River Ballad
The Chinese State Banquet

452
苍劲中的柔和美
渔人码头时尚鱼火锅
Vigorous in the Soft Beauty
Fisherman's Wharf fashion fish Hot pot

456
空谷生幽兰
妙香素食馆
Orchard in the Valley
Miao Xiang Vegetarian Diet

462
东方韵味
三义和酒楼
Ink painting of Dongqian Lake
A High-end Private Club in Yue Fu

468
诠释四合院文化 唐会
Interpretation of Courtyard Culture
Tang Club

474
墨韵 平潭某餐饮会所
Ink Rhyme   A Dining Club in Pingtan

480
舌尖上的岭南
舌尖岭南连锁餐厅
The Tongue of South of the Ridges
Tongue South of Five Ridges Chain Restaurant

484
清歌一曲月如霜 台门酒会
Sing Like Cream
The Stage Door Wine Party

第五章 办公空间
The Fifth Chapter : Offices Space

490
禅意的平衡之美
郑树芬设计办公室
Zen Beauty of Balance
Simon Chong Design Office

500
一行一世界 一静一禅心
无印良品空间设计办公室
A world of Quiet a Zen
Muji Design Office

504
龙•鼎
某金融投资公司办公区设计
Dragon• Ding
Design of a Financial Investment Company Office Area

508
素色古味 设计师的工作室
Plain Ancient Flavor
The Designer's Studio

The First Chapter
Club Space

第一章
会所空间

中华文化盛宴
臻会所
China Cultural Feast
Excellence Club

设计公司：SCD(香港)郑树芬设计事务所	Design Company: Simon Chong Consultants Ltd.
设计师：郑树芬、杜恒、许志强	Designer: Simon Chong, Amy Du, Zhiqiang Xu
项目地点：深圳	Project Location: Shenzhen

沏茶、品香、读书、观花、赏画……文人墨客的生活令人无限向往，无论外面的世界是纷扰还是繁华，在这里，臻会所是一场关于中国文化盛宴的汇聚，一方净土总能使人心境如水、让人流连忘返。

臻会所是一家私人俱乐部，也是餐饮休闲空间，由知名商业地产商深国投置业在深圳中心区开发，由SCD郑树芬设计事务所打造而成。臻会所位于市区繁华路段嘉信茂购物中心内，紧邻山姆会员店，交通便利，地理位置优越。

设计师当初与甲方接触时，甲方给出的要求简单而复杂：现代中式、低调奢华，这是一个深奥的主题，设计师的创作带有浓郁的传统味道，方案设计长达半年，立刻得到了甲方的高度认可，有种"众里寻他千百度，那人却在灯火阑珊处"的感觉。设计师从设计创作到汇报，从材料选型到施工跟进，亲力亲为，把握设计过程的每一个重要节点和环节，对空间关系的深度解构和微妙细节的把握，无形中碰触着我们的心灵，使我们深深地被空间所营造的氛围感动。

Making tea, tasting fragrance, reading, watching flowers, appreciating paintings... life of men of literature and writing makes people yearn for. Whether the outside world is turmoil or prosperous, Pegasus Club converges the Chinese cultural feast. One pure land always makes people mind like water, and linger on.

Pegasus Club is a private club –cum– dining and relaxation place, which is developed by the well-known commercial estate SZITIC in the central area ofShenzhen, and built by SCD Zheng Shufen design firm. Pegasus Club is located in the urban downtown section Jiaxinmao shopping center, close to Sam's Club stores, with convenient transportation which is crowded.

When the designer contacted with the first party in the beginning, the first party gives simple and complex requirements: Modern Chinese, understated luxury, it is a profound theme. The creation is with a strong taste of tradition. The project is designed up to six months, and got the highly ratification immediately of the first party, a kind of "Find his congregation where thousands of degrees, when I look back, that people in the lights Bend "feeling. From creative to the report, from material selection to construction follow-up, designer all does it herself, and grasp each node and an important part of the design process. the depth of deconstruction spatial relations and careful mastery of subtle details touch our soul virtually and moved by the atmosphere created by the space deeply.

神韵 新中式空间设计典藏

神韵 新中式空间设计典藏

Romantic charm Collection of New Chinese Space Design

神韵 新中式空间设计典藏

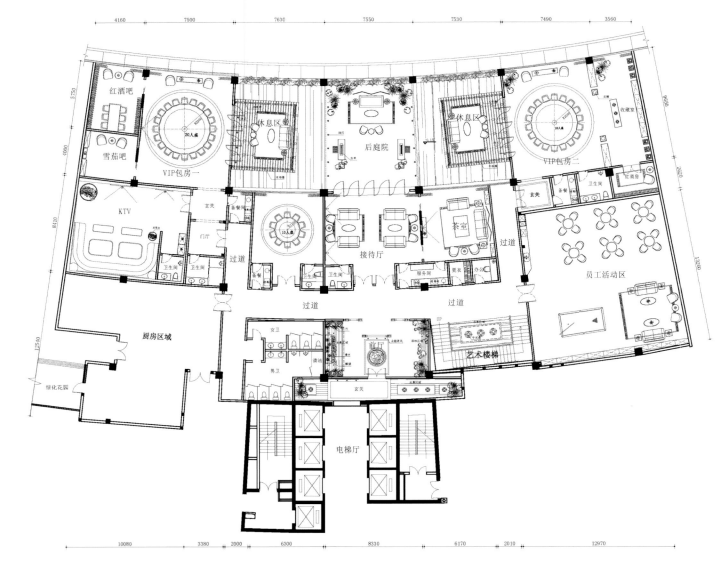

平面布置图

神韵 新中式空间设计典藏

中国红的现代演绎
融汇半岛会所
The Modern Interpretation of China Red
Integrate Peninsula Club

设计单位：重庆品辰装饰设计有限公司
主设计师：庞一飞、袁毅
项目面积：2200 ㎡
主要材料：木纹石、爵士白石材、黑色拉丝不锈钢、皮草等
开发商：融汇地产
摄影师：张起麟

Design Company: Chongqing Pin Chen Decoration Co., Ltd.
Main Designers : Pang Yifei, Yuan Yi
Project Area: 2200 ㎡
Main Materials: wood stone, jazz white stone, black brushed stainless steel, fur etc.
Developer: Integration Estate
Photographer: Zhang Qilin

"享受悠闲生活当然比享受奢侈生活便宜得多。要享受悠闲的生活只需要艺术家的一种性情，在一种全然悠闲的情绪中，去消遣一个闲暇无事的下午。"
——林语堂

会所原来是一个舶来品，意思是身份不凡人士聚会的场所，演变至今，已成为物业项目的配套设施之一。本案客户覆盖面广、功能齐备、社区客户多，作为社区活动的中心场所，以接待、聚会、休闲、运动和展示楼盘品质为主要目的。

公共区域的大厅挑高两层，空间通畅，没有丝毫的压抑感。内部空间的结构直白明了。简洁并富有自然肌理的大理石材质被大量运用，大面积的石材墙面构成沉稳的背景，塑造出一个干净整洁的形体，使空间得到释放和延伸。在这个有限的空间里，品辰设计用构成的手法将点与线的元素进行穿插，将其与空间体系有机组合，个性化的深色咖啡系与饱和度极高的中国红作为点缀，浓烈的色彩激发出饱满的情绪，营造出动感的活力氛围。镜面、玻璃及拉丝不锈钢的运用，不仅增加了时尚感，也提高了空间的通透度。

在陈设的选择上，品辰设计亦考虑到与整体空间风格的匹配，利落的线条和具有雕塑感的现代家具、马毛奶牛纹的靠垫、造型独特的树枝状吊饰，再加上其他艺术摆设品与绿植的点缀，呈现出细腻精致的陈设效果，使每个角度都充满了艺术、动感、跳跃的气息。

神韵 新中式空间设计典藏

" Of course, enjoying a leisurely life is much cheaper than the extravagant life. as long as with the artist's temperament, in a totally relaxed mood, go to entertain a leisure afternoon, you can enjoy a leisurely life." – Lin Yutang

The original club is a exotica. It is a meeting place for those who with extraordinary identity, and has evolved and become one of the supporting facilities of certified property projects. The customer coverage of this case is wide. The function is completed, and a lot of customers live in the community, as a central place for community activities, the main purpose is for the reception, party, leisure, sports and quality-showing of estate .

The hall of the public areas is two-floor high, open space, without the slightest sense of oppression. Structure of the internal space is straightforward and clear. Marble which is concise and rich with natural texture is extensive used. large area of stone walls form calm background, creating a clean and tidy shape, so that the space is released and extended. In this limited space, Pin Chen Design uses the technique with the elements of point and line, combined with its space system organically. Personalized dark coffee and high saturation of Chinese red as ornament, strong color inspires full emotion, creating a vibrant and dynamic atmosphere. The use of mirrors, glass and brushed stainless steel, increases the sense of fashion, and also improve the permeability of the space.

In the selecting of furnishings, Pin Chen Design also take the match of overall space style into account. Clean lines and sculptural modern furniture, horse hair cow stripe cushions, unique dendritic straps, together with other artistic furnishings and green plants, with delicate exquisite furnishings, make each angle fulling the atmosphere of art, movement, and jumping.

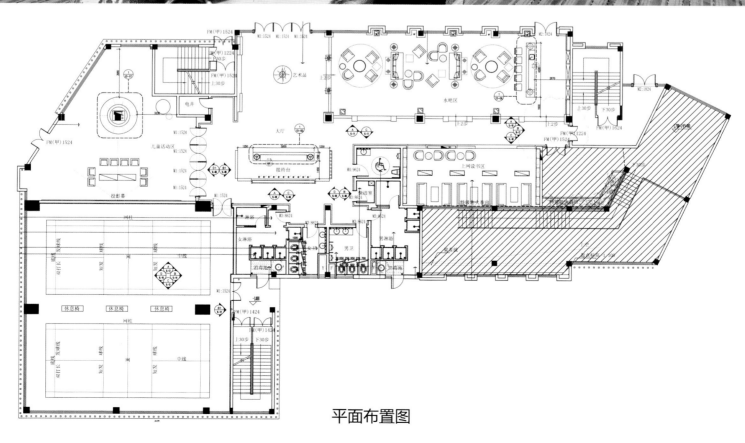

平面布置图

茶者，南方之嘉木也
金泰来名车茗茶坊
Tea, the Southern Fine Wood
Kingtailai Car & Tea Club

设计单位：台湾大易国际设计事业有限公司/邱春瑞设计师事务所	Design Company: Taiwan Dayi International Design Industry Co. Ltd./Qiu Chunrui Design Office
设计师：邱春瑞	Designer: Qiu Chunrui
项目地点：广东省深圳市南山区工业八路与科苑南路交汇处	Project location: Shenzhen City, Guangdong Province, eight Nanshan District Road, industrial and Ke Yuan Road Interchange
项目面积：400 ㎡	Project Area: 400 ㎡
摄影师：大斌室内摄影	Photographer: Da Bin Interior Photography
主要材料：银鼎灰大理石、印度木纹大理石、山西黑大理石、都市灰大理石、大花绿大理石、黑白根大理石、大花白大理石、仿古砖、青石板、钢化清玻璃、夹丝玻璃、艺术玻璃、黑镜、黑钛金、铜片、香槟金、乳胶漆、柚木饰面	Main Materials: Silver Pot Ash, India Wood, Shanxi Black, Grey, Green Stone City, Big Flower Green, Black and White Root, Big Grey, Antique Brick, Clear Tempered Glass, Wired Glass, Art Glass, Black Mirror, Black Titanium Gold, Copper, Champagne Gold, Latex Paint, Teak Finishes

"茶者，南方之嘉木也。"语出《茶经》。如今，江南茶馆如雨后春笋般出现，繁华的都市中，人们需要一个寄托心灵的场所，茶馆是最好的选择。茶道被看作是一种高雅的文化，茶坊可以满足人们审美、交流、养生保健等高层次的精神需要。茗茶坊是爱茶者的乐园，也是人们休息、消遣和交际的场所。

本设计作为名车养护中心的高端会所，是集休闲、销售于一体的现代商业空间。茶坊设有艺展台、展示区及六个泡茶坊，让每个进入到这里的人都能在品茶的同时还能欣赏茶艺、古筝表演，让人们能在品茗的过程中体会到一种全身心的放松，体验到心灵的净化和宁静。

本案将中国传统元素和粗犷材质相结合，将中国的传统文化更深层次地展现出来。在传统的茶馆风格设计的基础上，增加了多种自然的元素，采用原生态的石头及自然的木头等材料，让人们的心境直接回归到纯林的年代。流水型夸张的设计，不仅具有美的视觉冲击，更是艺术形式的展现。与之相呼应的小桥和石路设计，更是古朴，整块石材的运用，浑厚有力。大气的国画、优雅的水景、古朴的木格栅等古典造型元素与现代材质的完美结合，使整个设计形神统一。

荷花、荷叶以其高洁、素雅、出淤泥而不染的品质历来为文人墨客所称颂，茶馆对荷叶元素的合理运用，使人们在用餐时仿佛置身于一片荷塘之中，翠绿的荷叶、粉红的荷花、淡淡的荷香使人们心情舒畅。茶馆内的小木桥，桥下流水潺潺，荷叶随着水流不断摆动，荷花的清香扑鼻而来，让人立刻变得心情舒畅起来，使人仿佛置身于荷花的梦境之中。在喧哗的都市中构筑出一隅清香之地。琴音鸟鸣、水榭亭台、诗情画意，为客人提供了一份静如止水的休闲场所，瑟瑟的古筝、涓涓的流水，再配以荷花荷叶，给人带来一片喜气祥和的氛围。

设计师利用艺术吊顶将天花板凹凸不平的缺陷转变为与整个风格呈一体的木饰面。营造的是极富中国浪漫情调的生活空间，红木、青花瓷、紫砂茶壶及一些红木工艺品等都体现了浓郁的东方之美。带有现代中式纹样的吊灯，与暖色的射灯互相搭配，柜子里暗藏的灯带映衬着艺术陶瓷品，使人留连忘返。

"Tea, also South of the wood." The phrase "tea". Nowadays, such as bamboo shoots after a spring rain like Southern teahouse, the prosperous city, people need a spiritual sustenance place, teahouse is the best choice. The tea ceremony is regarded as a kind of elegant culture, tea house can satisfy people aesthetic, communication, health care and other high level spiritual needs. Tea tea is the love of tea of the park, but also the people rest, entertainment and communication.

This design as the cars maintenance center of high-end clubs, is the modern commercial space set leisure, sales in the integration of. Tea house is provided with a booth, art display area and six tea Fang, let each into the people here can also enjoy tea in tea, guzheng performance at the same time, so that people can feel a relaxed in the tea process, to experience the purification of the soul and calm.

The case will be China traditional elements and rough material combination, will China traditional culture deeply show. On the basis of the traditional teahouse style design, increases the variety of the natural elements, using the original natural stone, wood and other craft materials, let people's mood directly back to the simplicity of the age. Design of water type exaggerated, not only has beautiful visual impact, is to show more artistic form, bridges and road design echoes with the more simple, whole, using stone, powerful. The traditional Chinese painting color of the wind, Waterscape, wooden grille plain elegant classical model elements and the perfect combination of modern materials, so that the unity of the whole atmosphere design spirit.

Lotus, lotus leaf with its Gao Jie, simple and elegant, the silt but don't dye quality has always been for the praise of men of literature and writing, the rational use of the teahouse of lotus leaf element, make people at dinner as if in the midst of a lotus pond, green leaves, pink lotus, faint Hexiang makes people happy. The teahouse of the small wooden bridge, under the bridge water gurgling, lotus leaf with water constantly swinging, lotus fragrance from assail the nostrils, people immediately become happy together, people feel like being in the lotus dream. Build a corner fragrance in the hustle and bustle of the city. The sound of birdsong, pavilions, a quality suggestive of poetry or painting, provides a static, magnificent; ornate; fascinating leisure place for guests, the howling of the guzheng, trickling water, match again with lotus leaves, bring a happy and auspicious atmosphere.

Designers use art ceiling will defect of ceiling for whole style and uneven development of wood veneer. Build is highly China romantic living space, mahogany, blue and white porcelain, purple sand teapot and some mahogany crafts all reflect the rich Oriental beauty. With modern Chinese pattern chandeliers, spotlights with warm with each other, the closet hidden lamp store art ceramics authentic people linger.

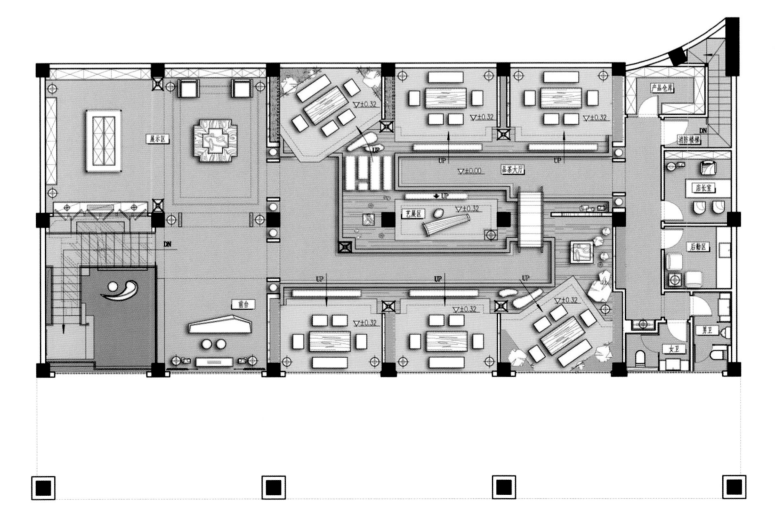

平面布置图

神韵 新中式空间设计典藏

东坡文化新诠释
汤泉会馆
Dongpo Cultural New Interpretation
Tangquan Hall

设计单位：台湾大易国际设计事业有限公司/邱春瑞设计师事务所	Design Company: Taiwan Dayi International Design Industry Co. Ltd./Qiu Chunrui Design Office
设计师：邱春瑞	Designer: Qiu Chunrui
项目地址：惠州	Project Location : Huizhou
项目面积：540 ㎡	Project Area: 540 ㎡
主要材料：砚石、灰麻石、黑金沙、银白龙、伯爵米黄、防雾银镜、茶镜、仿古铜、地毯、壁纸、木饰面、地板、乳胶漆、马赛克、亚克力	Main Materials: inkstone, gray granite, Black Sands, silver dragon, Earl beige, fog silver mirror, tea mirror, antique copper, carpet, wallpaper, wood finishes, flooring, paint, mosaic, acrylic
摄影师：大斌室内摄影	Photographer: Dabin Indoor Photography

茶馆内以素色为主调，粗糙的青石板和天然纹理的木地板厚实而流畅，仿佛划满了时间的痕迹，为整个空间带来一种大气磅礴的气势。茶馆以一种独特的姿态诠释着中式之美。

本案设计师将现代气息与东方禅意相融合，将空间演绎成一个优雅的品茗空间，设计以"茶"为引子，凝聚出整体的空间感，同时也向前延伸了空间体验。茶室的各个空间用木格隔成半透的空间，坐在包间内品香茗，心静则自凉。纵横结合更加脉络清晰，其复合性与包容性，赋予空间无限想象。呈现出细致优雅的空间氛围及简洁宽敞的空间感。

一楼的一间茶室朝北的方向全部以落地木窗代替墙面。屋外湖边的湖泊似乎成了茶室的一部分，俨然是一幅超大的立体水墨画。人在品茶的同时可以直面窗外的湖光天色。

苏东坡曾云："宁可食无肉，不可居无竹。"竹子常被赋予潇洒、高节、虚心的文化内涵，使观赏者通过观物而引申到意境，从而塑造一个清幽宁静的空间。设计师将一楼东、西两间茶室的墙面打空，墙与墙之间种上竹子，形成一幅浑然天成的水墨竹枝图。

二楼展示厅以"回归""内省"为出发点，展示出宁静、朴实的人文禅风，厅内仅摆着一件根雕，6 m 的墙面上则是用来投影展示。

三楼书房的设计以功能性为主。装修中考虑安静感、采光充足，这样设计有利于人们集中注意力。为达到这些效果，使用了色彩、照明、饰物等不同方式来营造。

设计师抛弃一切矫饰，尊重古建筑的原有语言，只保留事物最基本的元素，力求做到平淡致远。用最少的元素，（如：樱桃木、榉木、藤、竹等）来表达我们对苏东坡的敬意，东坡有词云："人间有味是清欢。"我们给大家呈现的也许是苏轼当年最喜欢的情境——素墙、黛柱、青地、白顶，在这种简逸的情境之中点缀着漏窗、竹帘、卧榻、古灯、幽兰、诗词、书法、绘画等，尤其在书画的设置上，设计师更是煞费苦心，尽一切所能搜集苏轼及与苏轼有关的传世书法、绘画作品，使用最接近原作的印刷方法复制，陈刊于室内及室外墙面，让近千年的东坡文化流淌在时空和空间之中。在这里，我们也许能够体味出当年以苏轼为首的文人雅士风云际会的畅意画面。

神韵 新中式空间设计典藏

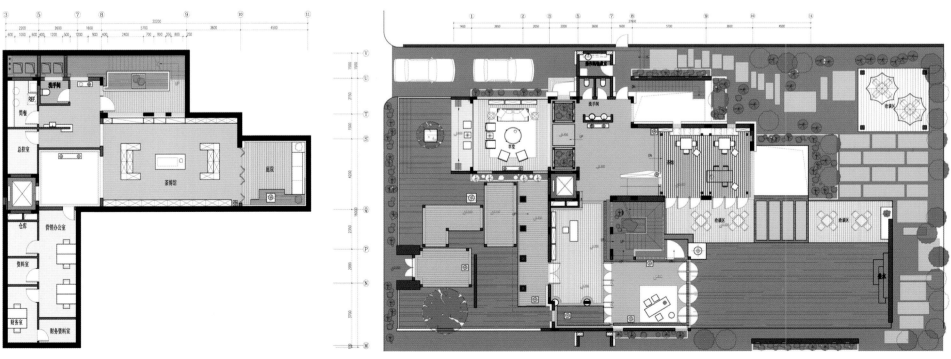

负一层平面图   一层平面图

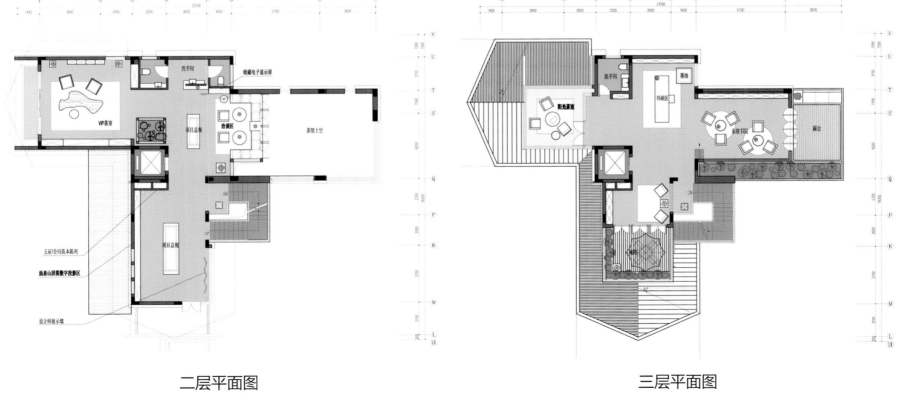

二层平面图　　　　　　　三层平面图

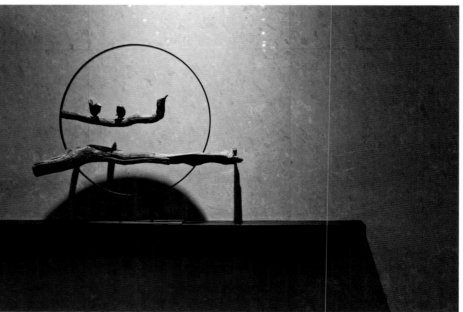

神韵 新中式空间设计典藏

The teahouse with plain tone. Rough quartzite and natural wood floors are thick and smooth, as if drawing full of traces of time, bring a majestic momentum for the entire space. Teahouse interpret Chinese beauty in a unique gesture.

The designer of this case kneads modern fashion and Oriental Zen, interpreting the space into an elegant tea room. "Tea" as primer of design gathers the overall sense of space and also extends forward the spatial experience. Each space of tearoom aare divided into semi– transparent space with wooden lattice. Sitting in the packages, sip a cup of tea, the calm nature is cool.Vertical and horizontal combination made context more clear, its complexation and inclusivity give space unlimited imagination, showing a delicate atmosphere and concise and spacious space.

Landing wooden windows of a tearoom facing north direction in the first floor all instead the walls.The outside lake seems to be part of the tearoom, just like a huge stereoscopic ink. While sipping tea, people could face the lake and sky outside the window.Su Dongpo was said: "I would rather eat no meat, can not live without bamboo". Bamboo is often given chic, high festival and modest cultural connotation, making the viewer through observing things then extended to the mood, thus shaping a quiet and serene space. Designer hits the two western and eastern tearooms' wall empty on the first floor, and plants bamboo between exterior, forming a natural bamboo ink figure naturally.

The second floor exhibition hall puts "return "and "introspection "as the starting point, demonstrating a quiet, simple Zen humanities. There is only stood one root in the hall. The six –meter wall is used for projection display.

The design of the third floor gives priority to functionality. Quiet sense of decoration, adequate lighting, it is good for concentration. To achieve these results, use different ways in color, lighting, accessories, etc.

Designer abandons all pretense and respect the original language of ancient buildings, only leaving the most basic elements of substance, and striving to achieve insipid. A minimum of elements (such as: cherry, beech wood, rattan, bamboo, etc.) express our respect for Su Dongpo. Su Dongpo once side: "The most flavored in the world is light joy"What we present perhaps Sushi favorite situation in those years – plain walls, Dai column, Green,white top.There are dotted with leaking window, bamboo, couch, old lamps, Orchids, poetry, calligraphy, painting, etc. in such simple situations, especially in the set of paintings. Designer is painstakingly make every effort to collect Sushi Sushi as well as related calligraphy collection and paintings. Using the most closest to original method to product, put in the indoor and outdoor walls, so that nearly a thousand years of Dongpo culture can flow in time space and space.Here, We might be able to savor the year when the literati headed by Su Shi enjoyed the joyful life.

神韵 新中式空间设计典藏

神韵 新中式空间设计典藏

古典中国的印象
北京龙潭湖九五书院
Classical China impression
Beijing Longtan Lake Paramount Chamber

设计单位：上海泷屋装饰设计有限公司	Design Company: Shanghai RID Co., Ltd.
设计总监：小川训央	Design Director : Ogawa Norio
设计师：佐佐木力	Designer: Sasaki Chikara
项目地点：北京市东城区左安门内大街20号	Project Location: Chongwen District of Beijing City, left anmen Avenue No. 20
项目面积：2650 ㎡	Project Area: 2650 ㎡
摄影师：贾方	Photographer: Jiafang

中国的历史性建筑物，作为受到保护的重要文化财产，在生动叙述着中国这个历史悠久的伟大国家的光辉文化。如何将这深远的寓意，变换成符合时代的设计呢？

这个设计，业主十分强调"古典中国的印象"。这对于身为外国人的我们是个难题。然而，我想正因为我们是外国人，才能演绎出新的、不同的美感。中国传统建筑物的主要特色体现在独特的石阶、顶棚和墙面装饰上。采用"在某个部分配合某种装饰"作为设计的基础，以此为范本，通过变换的方式和素材的叠加等方法进行设计。

在2650㎡的宽阔会馆内，风格多样的设计，使人们能够在不同的包厢欣赏到不一样的艺术感。比如说，顶棚上吊有艺术品的房间，强调突出了顶棚的高度，通过改变传统的吊顶装饰，来塑造室内氛围的变化。除此之外，通过在顶棚上贴木材，避免了"纯粹古典中式"的取向，还要特别注意不能设计成纯粹的现代中式风格。就我们的设计初衷而言，是在"古典"的设计基调中，感受到"当代"特有的时代感。

会馆内，也设置了可以摆放古董的展示空间，这不仅是店铺的设计，同时还能欣赏到中国的古典艺术。

"那是湖畔处若隐似浮的庄严姿容，朱漆的外表和豪华奢靡的内里，一瞥之间虽是古典之态，但细部却满是新中华风韵。"我认为这很好地表现了前述的设计理念。

Chinese palace and historic buildings, as an important cultural property is protected, in a lively account of glorious culture China this great country with a long history.How will this far-reaching meaning, transform into accord with the design?

In this design, the owners and emphasizes the importance of"classical Chinese impression".This is a problem for as foreigners, we. However, I thought becauseforeigners, in order to perform a new, different beauty. First of all, on the historic buildings China speaking, the main design, unique stone steps,ceiling and wall decoration. The "in one part with a decorative" asfoundation, and then as a template, by changing the way and the materialand its use for design.

In the composed of 2650 ㎡ of spacious hall, the design stylediversity, can enjoy different atmosphere in different boxes. For example, the ceiling hanging a art room, highlighting the ceiling height, by changing thelong existing China traditional art way, change to shape the indooratmosphere. In addition, through the wood on the ceiling, avoid "purelyclassical Chinese" orientation at the same time, we should also pay special attention to not designed for modern Chinese style. We designed ourspeaking, is in the "classical" this motif, also feel the "contemporary" specialsense.

Hall, also set can display antiques exhibition space, can enjoy not only thestore design, and Chinese classical art.

It is a solemn appearance if the implicit like floating lake. Is a red lacquerappearance and luxury luxury inside, a glance is the classical state, but the detail is full of new Chinese style. I think a good performance of the design concept of the.

纳西古民的院落生活
束河元年
Naxi Ancient Civilian Courtyard living
Shuhe First Year

设计单位：成都风上空间营造设计顾问有限公司	Design Company: Chengdu Fengshang Space Construction Design and Consulting Co, Ltd
设计师：王峰	Designer: Wang Feng
项目地址：云南丽江束河古镇	Project Location : Lijiang Shuhe
项目面积：3500 m²	Project Area : 3500 m²
主要材料：生态木、乳胶漆、水曲柳面板、中国黑石材	Main Materials: eco- wood, latex paint, Ash panels, Chinese black stone
摄影师：贾方	Photographer: Jia Fang

束河元年坐落于丽江束河古镇。设计理念来源于古镇古朴的风貌，以"院落生活"为基本形态，搭配富有民俗韵味、古意盎然的家具饰品。许多纳西古民生活与劳作的物件也重新萌生出美感，瓶、箱、笼、凳、椅以各自独有的造型，呈现出不同的视觉美感。

接待大厅中央纳西风格的立雕顶梁柱体现着古朴的纳西文化。纳西木雕是本案的设计灵魂，从餐饮区室外的连环木雕到接待大堂的顶梁柱，从走廊的立柱到客房的横梁，再到房门、号牌等，贯穿于整个空间中。为了体现这一传统装饰的精华之处，所有的装修材料、配饰都以传统经典材质为主。特殊烧制的青条砖、实木、硅藻泥、稻草板、铜艺灯等，还有当地的各种土生材料，比如自然荒石和老木头等更是院子里独一无二的风景。

The Shuhe first year is located in the ancient town – Shuhe Lijiang. The design concept derived from the quaint style of the ancient town with " courtyard life" as the basic form and furniture & accessory fulling of flavor of folk custom and ancient. Many working and living objects of ancient Naxi people also re-initiation of beauty. Bottles, boxes, cages, stools and chairs all show different visual aesthetics with their unique style.

The Naxi –style vertical carving pillar in reception hall embodies the ancient Naxi culture. Naxi woodcarving is the soul of the design of this case, which runs throughout the entire space from the outdoor interlinked woodcarving in the dining area to the pillar in reception hall, from the column in the corridor to the beam in the guest room, then to the door, license plate etc.. In order to match the essence of traditional decoration, all the decoration materials and accessories focus on traditional classic material. Special fired green brick, wood, diatom ooze, strawboard, copper lights etc., as well as a variety of local native materials such as natural stone and old wood, are the unique landscape in the yard.

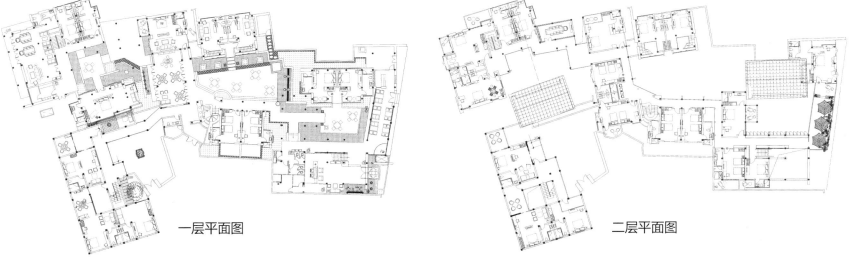

一层平面图　　　　　　　　二层平面图

闲坐庭前细品茗
顺义号茶艺馆
Drinking Tea before the Court
Shunyi Art Tea House

设计单位：深圳市盘石室内设计有限公司/吴文粒设计事务所	Design Company: Shenzhen City Panshi Max Interiors Co. Ltd., / Wu Wenli Design Office
设计师：吴文粒、陆伟英	Designer: Wu Wenli , Lu Weiying
项目地点：深圳梅林	Project location: Shenzhen Meilin
项目面积：500 ㎡	Project Area: 500 ㎡

中国茶文化源远流长，博大精深。中国茶文化发于神农，闻于鲁周公，兴盛在唐宋明清。中国自古以来便有以茶交友，品茶论道的传统。中国人饮茶，注重一个"品"字。"品茶"不但是鉴别茶的优劣，也带有神思遐想和领略饮茶情趣之意。在百忙之中泡上一壶浓茶，择雅静之处，自斟自饮，可以消除疲劳、涤烦益思、振奋精神，也可以细啜慢饮，达到美的享受，使精神世界升华到高尚的艺术境界。因此，这个品茶的环境就相当重要了，饮茶要求安静、清新、舒适、干净。中国园林世界闻名，山水风景更是不可胜数。利用园林或自然山水间，用木头做亭子、凳子，搭设茶室，给人一种诗情画意之感。

一个茶空间的完美呈现需要设计师具备深厚的茶文化底蕴，设计师通过对传统茶文化的认知，结合现代人的生活方式和审美形式，以自己的视角诠释中国悠久的传统文化精粹，演绎出具有东方哲学和现代生活美学于一体的茶饮休闲体验空间。

"偷得浮生闲半日，静坐庭前细品茗"，创造出梦境中的茶饮空间，看透人世的尘埃，浮浮沉沉。

Tea culture is well-established, broad and profound.China tea culture in Shen Nong, Wen Lu Zhou Gong, flourished in the Tang and Song dynasties of Ming and Qing dynasties. Chinese since they have friends to tea, tea on the traditional. Chinese tea, pay attention to a "product". "Tea" is not only to distinguish the quality of tea, but to meditate and taste the feeling of drinking tea. Bubble a pot of tea in his busy schedule, choose a quiet place, Zizhenziyin, can eliminate fatigue, di annoying extend, hearten spirit, can also drink it slowly, enjoyment of beauty, that the spirit of the world into a sublime realm of art. Therefore, the environment is very important for tea, tea requirements quiet, fresh, comfortable, clean. China gardens are well known in the world, the landscape is countless. Use of the natural landscape or garden, pavilion, wooden stool, erection teahouse, give a person a kind of a quality suggestive of poetry or painting. For people to rest, not interesting.

The perfect one tea space presents designers need to have the rich cultural heritage of tea, the designer through the traditional tea culture cognition, combined with the modern way of life and aesthetic form, with their own interpretation of China long traditional culture, interpretation of the tea leisure experience space with oriental philosophy and aesthetics in the integration of modern life.

"Stolen floating free half day, sit before the court of fine tea", to create a dream tea space, look across the air of dust, ups and downs.

神韵 新中式空间设计典藏

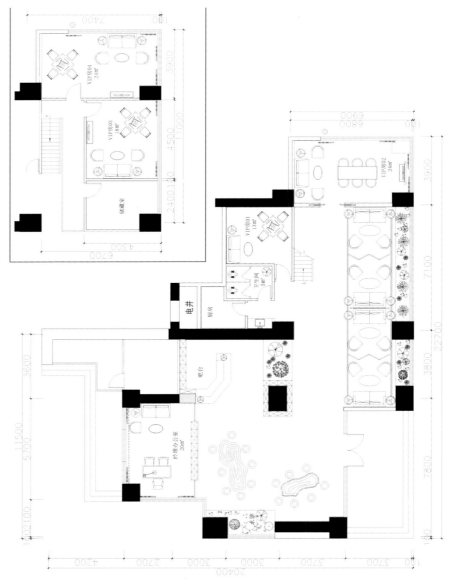

一、二层平面图

神韵 新中式空间设计典藏

八闽首府之幽幽岁月
聚春园驿馆

The Capital of Fujian in the Peaceful Years
Ju Chun Yuan Station

设计单位：福建国广一叶建筑装饰设计工程有限公司	Design Company: Fujian Guoguang Yiye Architectural Decoration Engineering co. Ltd
方案审定：叶斌	Project Validation : Ye Bin
设计师：金舒扬、刘国铭、陈剑英、李宏、王其飞、蔡加泉、张慧晶、余峰	Designer: Jin Shuyang, Liu Guoming, Chen Jianying, Li Hong, Wang Qifei, Cai Jiaquan, Zhang Huijing, Yu Feng
项目地址：福州三坊七巷	Project Location : Seventh Alley Three Lane Fuzhou
项目面积：4000 ㎡	Project Area: 4000 ㎡
主要材料：青石板、仿古砖、乳胶漆、杉木做旧	Main Materials : quartzite, antique tiles, latex paint, old fir

作为八闽首府的福州，自闽越王无诸建治城至今已2200余年，创造了壳丘头文化、昙石山文化等富有地域特色的区域文化，养育了林则徐、严复、沈葆桢等一批近现代名人。本案坐落在具有悠久文化底蕴的建筑群落——三坊七巷中。设计师秉承设计要立足在坊巷整体格局特点之上，提取再整合，取之再跃之的设计理念。贯穿将设计融于自然，融于所处的建筑群落之中，使中式空间富有层次，韵味十足。

驿馆在古代是供传递官府文书的人途中更换马匹或休息、住宿的地方。我们在设计驿馆的时候理解的当代驿馆，应该不似别墅或豪华宾馆那么奢华，布局要讲究中国风水，体现中式的韵律感。本案的设计既大气优雅，又内敛简约，彰显出中国智慧和东方精神。

该项目的设计还有清末明初的建筑身影，在设计时汲取了19世纪末期建筑的特点，并融合了一些现代元素，其中一些空间采用了东情西韵的调子来诠释，让中国式的大气沉稳与西式的柔美优雅共处一室，使整个空间散发独特的韵味。总之，在这里可居、可观、可游、可赏，在这里，随心、随性、随情、随景。

As the capital city of Fuzhou, Fujian has a history of 2200 years since Wuchu built no such rule Zhi city, creating regional cultural full of regional features such as Keqiutou culture, Tanshishan culture etc. and raising Lin Tse-hsu, Yan Fu, Shen Pao-chen and a number of modern celebrity.This case is located in the architectural community of rich cultural – Seventh Alley Three Lane. Designers adhere to the idea that design should be based on the characteristics of the overall lane-alley pattern, extract and then integrate. Proceed from it, but better than it. Put the design into the natural, and into the building community. Let Chinese style be rich with the layers and full of flavor.

Courier station is a place for people who transfer official documents replace horses or take a rest in ancient. When we design the courier station, the contemporary courier station should not seem so sumptuous like villas or luxury hotels. The layout should pay attention to Chinese feng shui and reflect Chinese sense of rhythm. The design of this case is not only grace and atmospheric, but also simplified and conservative, which highlights China wisdom and Eastern spirit.

There are architectural figure of late Qing and early Ming Dynasty in the design of this project. The design draws architecture characteristics from the late 19th century, and blends some modern elements. Some of which adopts the tune of East West rhyme, so that the Chinese mellow atmosphere fuses Western elegance together, which makes the entire space exude a unique lingering charm. Anyway, here is livable, attractive and enjoy. Here is arbitrary, casual and emotional.

神韵 新中式空间设计典藏

神韵 新中式空间设计典藏

神韵 新中式空间设计典藏

神韵 新中式空间设计典藏

诗情画意醉东方
福临门会所

Oriental Poetry Drunk
Fook Lam Moon Club

设计单位：品尚空间装饰	Design Company: Pinshang space decoration
设计师：林崇	Designer: Lin Chong
摄影师：李玲玉	Photographer: Li Lingyu

福临门会所的大门被塑造成古时私家宅院的样式，静静地坐落在繁华的都市中。门前高挂的大红灯笼、朱黑的匾额、烫金的题词，两侧石鼓竖立，绿植郁郁葱葱，让人不禁想象门后那个神秘的空间。推开厚重的镶着如意铜扣的红木大门，眼前不由一亮。一扇仿古雕花窗格木屏风半遮视线，高达两层楼的挑高空间有着天井一般的效果。这里悬挂的大型吊灯，凸显大气风范。墙面、地板均使用天然石材装饰，黑色、米白色、金沙色拼贴出的纹样干净沉稳，并孕育着空间不可或缺的气度。

一楼宽阔的空间是半开放式的用餐区，沿着窗户的四周布置上舒适的沙发、桌椅。此外，还有轻垂的竹帘和红木屏风隔断，既能装饰，也保证了空间的私密性。沿着黑色大理石楼梯缓步至二层，墙上实木窗格中装饰的青花瓷盘创意独到。它在深沉的色彩里投下一抹清明，一楼休闲平台的山水画卷屏风被灯光映衬的灵动、高贵，既虚实结合，又充分提升了空间的艺术气息。二楼大大小小的包厢各有特色，或雍容典雅、或清丽脱俗。中庭两侧的包厢是设计师重点打造的部分，用窗棂面对中庭，让在此包厢中用餐的客人，能透过中式窗花的窗子，看见贯穿的一二楼中庭空间，别有一番意境。

会所内的陈设饰品皆是设计师精心挑选，极为细腻精致。挂画、花瓶、小工艺品等都散发着古韵和优雅的内涵。空间以简约的造型为基础，添加了窗棂、方格造型等中式元素，使得空间大而不空、厚而不重。福临门会所用温暖的古建筑替代了冰冷的现代建筑，安静怡人的空间环境让人倍感舒心，深藏的历史韵味和文化气息值得细细品味，高品质的空间体验让人沉醉其中。

The gate of Fook Lam Moon Chamber was fashioned into the style of ancient private house, situated quietly in the bustling noisy metropolis. In front of the gate, red lanterns hanging, black plaque, gilt inscription, erected stone drums on both sides and lush green plants all make people can not help looking at the mysterious space hiding behind the door. Open the heavy mahogany gates with Ruyi copper buckle, eyes could not help lighting up. An archaized carved wooden screen covers the view. The high ceiling up to two-story has the effect of patio. The large hanging droplight highlights atmospheric demeanour. Walls and floors are decorated with natural stone. Patterns spliced by black, creamy white and golden sands color is clean and calm and pregnant with integral presence of space.

Wide space of the first floor is semi-open dining area, arrange comfortable sofa and chairs along windows. In addition, there are bamboo curtain and mahogany screen, which are both for decoration and ensure the privacy of the space. Step into the second floor along the black marble staircase, Blue and white porcelain in the wood pane of the wall is Creative and original as if a pearl among deep color. Landscape screen in the first floor leisure platform is smart and noble against the light, virtuality and reality combing, and fully enhance the artistic space. Large and small balconies on the second floor have their own characteristics, or graceful and elegant, or elegant and refined. Boxes on both sides of the atrium are the part designer focused on. Facing the atrium with window frames, dinner guests in this box can see atrium space between the first and second floor through the Chinese grilles window, which have a specific taste.

Furnishings in the chamber are exquisite and delicate and carefully selected by designer. One painting, a vase, or a small craft exudes ancient and elegant connotation. Space is based on simple shapes, and adds window frames, box modeling and other Chinese elements, so that space is large but not empty, thick but not heavy. Fook Lam Moon Chamber's warm ancient architectures replace the icy modern buildings. Quiet and pleasant atmosphere makes people feel comfortable. Strong historical charm and cultural ambiance are worth to taste carefully. And high-quality feast draws people to indulge in it.

神韵 新中式空间设计典藏

一层平面图

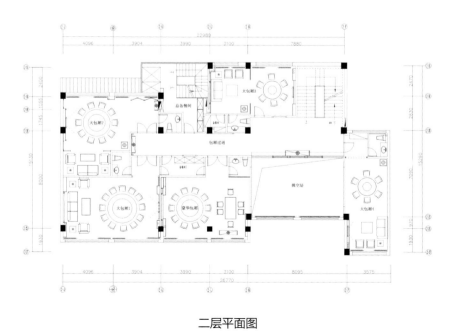

二层平面图

神韵 新中式空间设计典藏

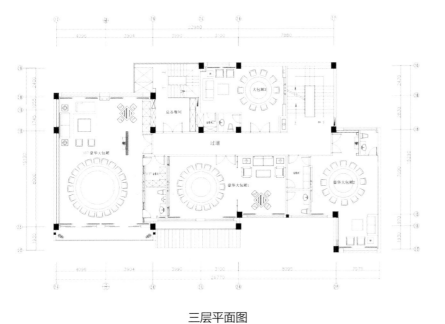

三层平面图

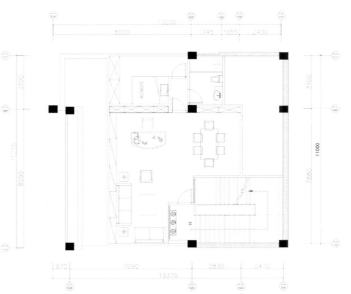

四层平面图

神韵 新中式空间设计典藏

神韵 新中式空间设计典藏

神韵 新中式空间设计典藏

阿弥陀佛世界
泰·自然养生会所
Amitabha World
Thai · Natural Health Club

设计单位：福州品川装饰设计有限公司
设计师：郭继
项目地点：福州

Design Company: Fuzhou Pinchuan Decoration Design Co., Ltd.
Designer: Guo Ji
Project Location: Fuzhou

这家泰式养生会所，汇集了泰式设计风格的精华。浓郁的色彩、葱郁的绿木、随处可见的草编和木质家具摆件，演绎出一场华丽的异域风情。设计师从材质入手，大量使用草编壁纸、橡木饰面、通花、文化石、实木地板、仿古砖等材料，力求还原泰国的本土风貌，带给客户最原汁原味的泰式服务。

楼正厅以辉煌的金色和热烈的红色吸引住客人的眼球，给人目不暇接的绚烂感受。正厅中央金黄色庙宇设计的空间散发着浓烈的泰式风情，如同将清迈的庙宇搬到了这里，让人不禁想入庙一观。大红色的墙面使用暗纹软包装饰而成，大片的红色与金色相互辉映，将正厅打造成了金碧辉煌的泰国宫殿，彰显出会所不凡的品味和地位。

步入二楼，是供客户休息、闲淡的茶话厅。茶话厅陈设古朴，以原木色和金色为主，将泰式风格中崇尚自然、古典质朴的一面呈现眼前。墙面以黑金色的草编壁纸铺就，再加上金黄色的荷叶壁挂，展现出泰式风格对金色的青睐。

会所的长廊也是一大特色，来到拐角处，宽阔的走廊灯光熠熠，一侧墙面上铺陈着泰国古风貌的壁画，再以文化石和热带植物点缀，竟有逼真的效果凸显出来。往里走，长廊越发幽深，晕黄的壁灯氤氲出朦胧梦幻的情境。黑金色的草编墙纸在灯光的映照下，焕发着低调的奢华。两旁身姿妖娆的木头摆件，也在地灯的照射下，散发着金子般的光芒，将幽长的走廊点缀得更加华丽。

按摩足浴房仍然以木色为主，墙面以素色壁纸铺就，简单而大方。墙上挂着泰式油画和泰式织物挂件，加上镂空的木质门窗，没有多余的点缀，就将泰式按摩间的大气质朴展现出来。

This Thai health club brings together the essence of Thai design style. Rich colors, lush green trees, straws which can be saw everywhere and wooden furniture and decorations deduce gorgeous exotic style. Starting from the materials, designers extensively use straw wallpaper, oak veneer , flower , cultural stone , wood floors, antique tiles and other materials , strive to restore native Thai style and bring customers the authentic Thai service .

The brilliant gold and warm red in the main hall catch the eye of the guests, giving people a feeling of dizzying gorgeous. The golden temple designed space in the central main hall exudes a strong Thai style as the temples in Chiengmai were moved here, and people can not help to take a look. The

bright red wall was decorated by dark fringe flexible packs, Red and gold mutual reflect, which builds the main hall into resplendent and magnificent palace of Thailand and highlights the extraordinary taste and status of the club.

The second floor is a tee room for customers to take a rest and chat. The tea room's furnishings of quaint give priority to the original wood color and gold, and show us Thai style's advocate natural and classical plain. The wall was paved with black golden straw wallpaper and golden lotus leaf wall hanging which shows the favor of Thai style on the golden.

The long corridor of the club is also a major feature. Come around the corner, lights are sparkling in the wide hallways. There lays ancient Thai style murals on the other side wall with cultural stone and tropical plants interspersing next, which emerge the photorealistic effects.

Go inside, the long corridor is increasingly deep. The dusky wall lamps enshroud hazy dream situations. Black and golden straw wallpapers glow with understated luxury against the lights. The enchanting wood ornaments on both sides also exude golden light. The quiet long hallway is decorated even more gorgeous.

Massage and foot bath room is still mainly with wood color. The wall paved with plain wallpaper, simply and elegantly. Thai oil painting and Thai textile accessories are hung on the wall with hollow wooden doors and windows and no redundant ornament, which reveal the rustic atmosphere of Thai massage store.

神韵 新中式空间设计典藏

神韵 新中式空间设计典藏

神韵 新中式空间设计典藏

婉约徽派，中国画里的乡村
北京寿州大饭店SPA养生会所

Graceful Hui Style, Chinese Painting inside the Country
SPA Health Club of Beijing Shouzhou Grand Hotel

设计单位：许建国建筑室内装饰设计有限公司	Design Company: Architectural Interior Design Co., Ltd. Xu Jianguo
设计师：许建国	Designer: Xu Jianguo
项目地点：北京	Project Location: Beijing
项目面积：1500 ㎡	Project Area: 1500 ㎡
主要材料：意大利木纹石、水曲柳肌理板、仿古砖、原木、皮革	Main Materials: Italian wood stone, ash texture plates, antique tiles, log, leather

青砖门罩、石雕漏窗、木雕檐柱，如诗般美丽的徽派设计，总能让人获得一丝熟悉的感动和文化的美感，木格窗、满顶床、蓑衣、荷花的出现，则将这份美丽演绎得朴素而又生动。本案从徽派及后现代设计着手，营造出一个文化氛围浓厚的中式会所，它散发着古朴、端庄的东方韵味。整个会所以内敛、沉稳的棕色作主调，配以淡黄色灯光的烘托，更显包容与亲和感，在感受到中国传统文化气息的同时，也不显得压抑。

门口窗格花纹做装饰的隔断，镂空的设计既满足了功能区分的需要，又加强了区域间的交流；通过一池荷塘造景、几片荷叶踏步，推开一扇白色的木格栅门，进入一个古色古香的接待区域。前台桌子旁的灯笼造型座灯玲珑精致，散发着柔和静谧的光线。背景墙是一幅"残荷听雨"图，幽静凄美。而与前台相对的位置放着一张木雕大床，厚重而贵气。一幅清朝人物装饰画静静地挂在那里，浓郁的文化气息弥漫开来。昏黄的灯光透过白色木格栅吊顶投射进来，光影交错中更显古朴静谧。

Blue brick door shelter, stone carving leaking window, wood carving Anta. The poetic beauty Anhui design can always give people a trace of familiar touch and the beauty culture. The appear of wooden lattice windows, full top beds, rain clothes and lotus deduce this beauty rustically and vividly. This case start from Anhui style and post-modern design, creating a strong cultural atmosphere of the Chinese club which exudes a quaint, dignified oriental charm. The homophony of whole club is introverted and calm brown tone with the contrast of yellow light, which highlights the tolerant and affinity. There is no inhibition when we feeling the flavor of traditional Chinese culture.

The separation decorated windowpane pattern at the door and hollow design can not only meet the functional distinction needs, but also strengthen the exchanges between regions. Landscape through a lotus pond, circle in a few slices of lotus leaves, open a white wooden grill door, step into a quaint reception area. Lantern style lights beside the reception table are exquisitely exquisite emitting soft and peaceful light. The backdrop wall is a pair of "Withered Lotus" map, quiet and poignant. There is a carved wooden bed at the relative position of the front, heavy and noble. A decorative Qing figure painting hung in there quietly with rich culture flavor spreading out. dim light shoot in through the white wooden grille ceiling, simple and quiet.

神韵 新中式空间设计典藏

神韵 新中式空间设计典藏

神韵 新中式空间设计典藏

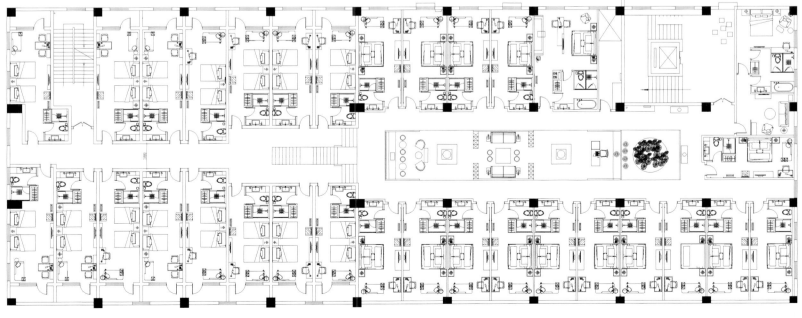

平面布置图

巴风蜀韵
成都天府高尔夫球会所
Bashu Charm
Chengdu Tianfu Golf Chamber

设计单位：美国圣拓建筑工程设计有限公司	Design Company: S.T.I.D.Group
设计师：林振中	Designer: Lin Zhenzhong
项目地址：成都青羊区	Project Location : Chengdu Qingyang District
项目面积：4500 ㎡	Project Area: 4500 ㎡
主要材料：灰砖、柚木、马来漆、椰壳	Main Materials : gray bricks, teakwood, Malay paint, coconut shell

本案地处成都青羊区，在设计之初，设计师走访了成都的多个古迹遗存，试图更多地感受四川的人文气质，了解四川传统建筑的内涵。成都平原孕育了与中原文化有别的蜀文化，这是地方特色浓厚的土著文化。

本案从风格上是一个休闲度假的中式会所，设计师注重具有对称秩序的空间布局，拥有优雅的流线，且注重借景。在材料使用上，设计师采用灰砖、柚木、马来漆、椰壳等，目的是营造出一个与众不同的空间氛围。

This case is located in Chengdu Qingyang District. In the beginning of the design , the designers visited a number of relics of Chengdu , trying to experience more humanistic qualities of Sichuan and understand the meaning of traditional architecture. Chengdu Plain gave birth to the Shu culture different from the Central Plains culture, which is indigenous culture with strong local characteristics.

From the style of this case, it is a Chinese leisure chamber. designer focuses on symmetrical spatial layout , elegant streamline and borrowed scenery. In the using of materials, the designer use gray bricks, teakwood, Malay paint, coconut shell etc.. to create a unique space atmosphere .

神韵 新中式空间设计典藏

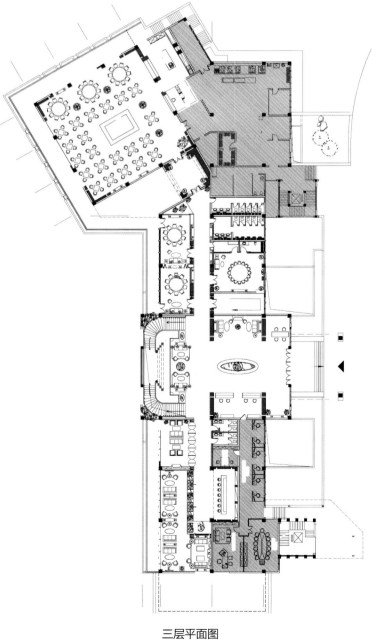

三层平面图

诗意江南
某茶楼设计
Poetic Jiangnan
A Teahouse Design

设计单位：美迪装饰大宅设计院	Design Company: Meidi Decoration Mansion Design Institute
设计师：赵益平	Designer: Zhao Yiping
项目面积：1500 ㎡	Project Area: 1500 ㎡
主要材料：木饰面、青麻石、墙绘、裱纸工艺等	Main Materials: Wood veneer, black granite, wall painting, paper crafts table

本案位于繁华闹市的一隅，投资方拟造一所以茶会友之所。商业定性为会所式，营业模式预采用会员制。目的是营造出一个氛围宁静、隐秘的居所，为当代精英提供一个会友、谈判、规划合作，以及娱乐的场所。

定案时，以"汇"字为中心贯穿空间，为了强调空间低调、含蓄的氛围，又融入了江南建筑的元素和中国传统文化的神意。在大厅部分，江南建筑中的廊柱运用其中——白色的墙、青色的砖，曲桥通幽、水映青莲……它们提炼了空间的气质，让消费的人群彻底洗涤世俗凡尘，回归到安静的、低调的江南文化之中。孔明灯的造型灯遍布其间，寓意吉祥和幸福。

水在中国人心中是吉祥如意的象征，而水井则是聚水之物。在长沙，白沙古井为这座古城蕴藏了不可多得的文化内涵。食品区中形似水井的顶棚突出"汇"的精髓及聚水为财的寓意，同时也强调了水在中国人心中的地位。在贵宾区，中国传统窗格纹样围合的走廊流露出了VIP空间的高雅定位。

整个空间采用了大量的木制品，形似深宫大院的柱廊，在数量上完成了"汇"的主题呼应，沉着稳重。当然运用在建筑中，不可避免地会让空间有所生硬，于是大量的墙绘应运而生，不仅丰富了空间的色彩，也增强了空间的艺术和唯美感。

This case is located in a corner of bustling downtown investor intends to build a tea club that making friends with others through tea.Designer determines the commercial nature as a club, intends to adopt the membership as the operating pattern and plan to create a quiet and hidden residence, so to provide a place that making friends, negotiating, planning cooperation and having fun for contemporary elites.

When establishing the scheme,the designer made the "Hui" word as the main object throughout the space.Meanwhile in order to emphasize low-key, implicative atmosphere of the space integrate into architectural elements of South of the Yangtze River and traditional Chinese culture spirit.In the hall, designer adopted the column building elements of architecture of South of the Yangtze River— white walls, blue bricks.Curved bridge goes through quiet and the water reflected the violet ... They refined the temperament of space so that the consumers thoroughly get rid of the secular world and return to the quiet and low-key culture of South of the Yangtze River.The models of the Kong Ming lantern can be found everywhere and they mean auspicious and blessing.

Water is the symbol of good luck and happiness in China and wells are the places to get gather water.In Changsha,Ancient Well of White Sand contains rare cultural connotation for the city.In the food area,shape like well ceiling expresses the essence of "Hui" and the implied meaning of gathering water as wealth,but also emphasizes the importance of water in the Chinese heart. In the VIP area, China traditional pane design enclosed corridor naturally show the fixed position of elegant of VIP space.

Adopting a plenty of wood products in the whole space.There is the shape like the palace courtyard colonnade.In quantity,they completed the "Hui" echoes the theme and it is calm and stable.Certainly, the use of architectural language which will inevitably somewhat makes the space dull. So there are a large number of wall paintings emerged, which not only enriched the colors of the space but also enhances the artistic and aesthetic sense of space.

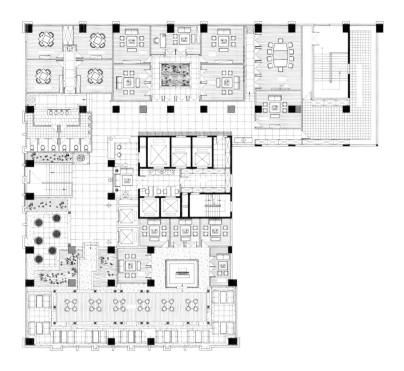

平面布置图

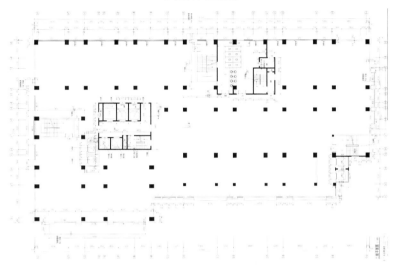

原始平面图

方圆之道
沁心轩茶会所
The Truth in Square and Circle
Refreshing Hin Tea Club

设计公司：福州中和设计事务所	Design Company: Fuzhou Zhonghe Design Firm
设计师：范敏强、陈锐峰	Designers: Fan Minqiang, Chen Ruifeng
项目地点：福州	Project Location: Fuzhou
项目面积：99 ㎡	Project Area: 99 ㎡
主要材料：水泥板、瓦片、茶叶盒、灰镜	Main Materials: Cement board, Tile, Tea box, Grey mirror

本案是一个面积不大的茶会所，自然和朴素是这个空间的重要特点，然而今天的审美和主流思维已经远离了那个传统的年代。于是设计师重新解构了中国文化的代表性元素，从色彩中提炼出黑和白，从形态中提炼出方和圆，从氛围提炼出闹和静，最终塑造出一个精神需求与物质享受相融合的禅意空间。

This case is a small area of tea clubs, nature and simplicity are important standards to this space. However, today's aesthetic and mainstream thinking has been away from the traditional era. So designers re-deconstructed representative elements of Chinese culture. They extracted black and white from all colors, extracted square and circle from all shapes, and extracted jollification and peace the atmosphere. Ultimately create a space of zen that combined spiritual needs and material pleasures together.

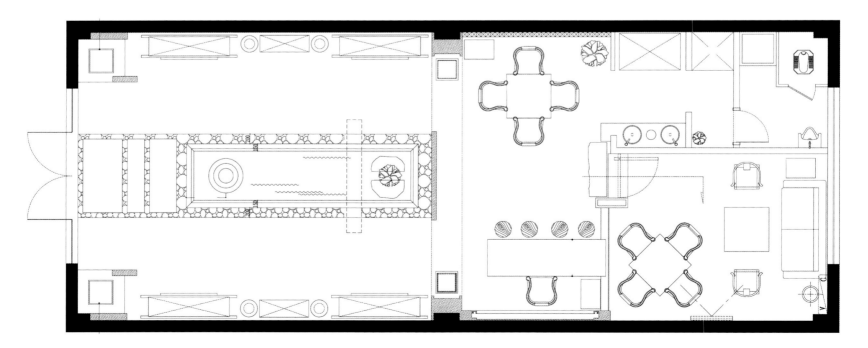

平面布置图

神韵 新中式空间设计典藏

印象茶文化
某茶楼设计
Impression Tea Culture
A Teahouse Design

设计单位：福州合诚环境艺术装饰有限公司	Design Company: Fuzhou Honest/Hecheng Environmental Art Decoration Co., Ltd.
设计师：陈家雄	Designer: Chen Jiaxiong
项目地点：福州	Project Location: Fuzhou
项目面积：120㎡	Project Area: 120㎡

本案设计以点、线、面为主，用黑、白色系来划分空间。入门处墙面镶嵌的茶壶如鱼儿般跃入二楼悬浮的展示架中，从不同角度、不同界面展示出它的形态，虚实互换，体现出设计的主题，同时也与茶文化相呼应。

一楼的接待台，像地面石材被翻起一样，显得生动而有立体感，与地面融为一体。一楼挑高的空间使得空间有了层次感。空间中多处运用了镜子，通过镜子的映射产生了虚实的空间结构，使空间更有层次，并且延伸感更强，同时也体现出了设计的主题。

This case is based point , line and surface on design and divides the space using black and white color scheme. The teapot inlaid into the wall at the entrance looks like a fish swimming into suspended display in the second floor and demonstrates its form from different angles and different interfaces. Swap generalization for concretization. It reflects the theme of the design and coincides with the tea culture at the same time.

The reception in the first floor is vivid and stereoscopic as the ground stone was turned up and integrated with it. The high-ceilinged space in the first floor let the space have administrative levels feeling. Multiple mirrors are used in the space. Though mirrors' mapping, virtual and real spatial structures are produced making the space more structured and extended, and reflecting the design theme at the same time.

神韵 新中式空间设计典藏

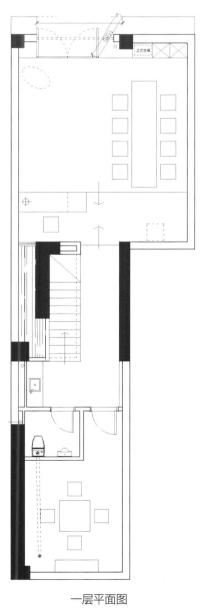

一层平面图

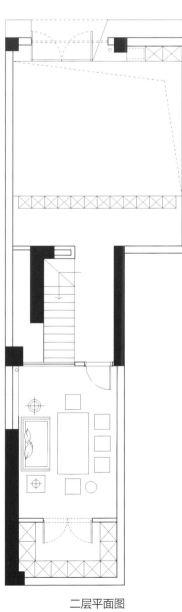

二层平面图

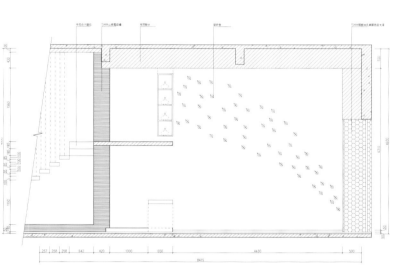

立面图1

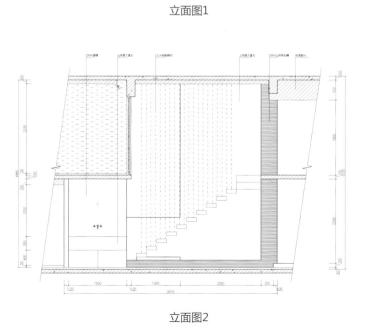

立面图2

神韵 新中式空间设计典藏

神韵 新中式空间设计典藏

水墨画乡
梅林阁会馆

Ink and Wash Paintings
Plum Pavilion

设计公司：合肥许建国建筑室内装饰设计有限公司	Design Company: China (Hefei) Xu Jianguo Architectural Interior Decoration Design Co., Ltd
主设计师：许建国	Chief Designer: Xu Jianguo
参与设计师：陈涛、程迎亚	Co-designer: Chen Tao, Cheng Yingya
项目地点：合肥	Project Location: Hefei
项目面积：260 ㎡	Project Area: 260 ㎡
主要材料：古木纹饰面板、小青砖、芝麻黑石材、仿古板	Main Materials: The ancient wood decorative panel, blue brick, sesame black stone, antique board
摄影师：吴辉	Photographer: Wuhui

本案位于合肥市黄山路，是与徽派文化一脉相承的主街。本案打破了传统徽派建筑的特点，选择具有浓厚茶文化底蕴的徽派风格来彰显本案特色，让人享受一份放松、优雅的环境，细细体会徽州茶文化的精髓，是本案设计的主旨。本案的外观运用了马头墙的形式排列，可以增强徽文化的印象，与传统文化相得益彰。

本案一楼是茶叶销售区，二楼是品茶区。门厅运用了书架式的隔断，减少外部环境对内部的影响，一楼分为前厅接待区、体验区、休闲景观区、茶叶展示区。茶叶展区有序地摆放着茶产品，展区四周有循环通道，方便顾客的流动。一楼景观区有古琴、书卷架、观音、假山水景，让人感受到一份平静、朴素、平和、自然的空间氛围。设计师把人造天井运用于本案中，其间的假山水景，巧妙地连接一、二两层楼，空间通透，采光效果好，二楼的顾客可以围绕天井欣赏一楼的布景，鹤与流水的造景相映成趣，给人一种回归自然的感觉。二楼饮茶区分服务区、休闲区、书画区、卧榻区、功能齐全，以满足不同客人的需求。

The case is located in Huangshan road, Heifei city and is the main street that comes down in one continuous line with Huizhou culture. A higher level of the surrounding crowd with it, breaking the traditional Huizhou architecture features, choosing a strong tea culture to highlight the case of Anhui style characteristics. The gist of the case design is that we can enjoy the relaxed, elegant environment and feel the essence of Huizhou tea culture. Arrangement of Ma Tau wall form was applied to the appearance of the case, which can enhance the impression of Huizhou culture and it complemented traditional culture very well.

The first floor of the case is the tea sales area and the second floor is tea tastes area. Bookshelf type partition was applied to hallway, which can reduce the impact of the external environment on the internal. The first floor is divided into the lobby n area, experience area, leisure landscape area, tea exhibition area. Tea exhibition area orderly placed tea products. Exhibition area was surrounded by circulation channel, which can facilitate customers to walk and to select products. Open space and north-south permeability and good lighting effects can let customers who on the second floor enjoy the scenery around the light. Crane and running water landscaping gain by contrast, giving people a feeling of returning to nature. Tea area of the second floor is divided into service area, leisure area, painting area and lounge area. Complete function can meet the different needs of the guests.

神韵 新中式空间设计典藏

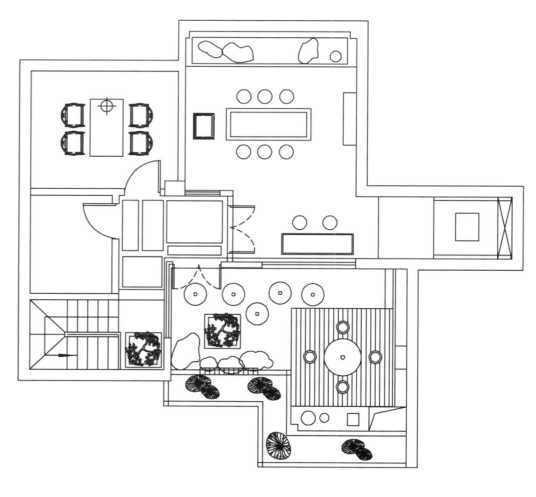

阁楼平面图

神韵 新中式空间设计典藏

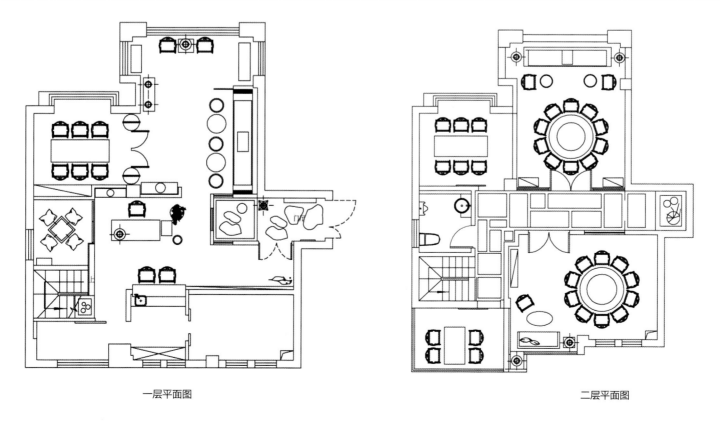

一层平面图　　　　二层平面图

神韵 新中式空间设计典藏

琵琶行
水墨江南会所
Song of the Pipa
Ink Jiangnan Club

设计单位：朱回瀚设计顾问工程（香港）有限公司	Design Company: Zhu Huihan Design Consultants (Hongkong) Co., Ltd.
设计师：朱回瀚	Designer: Zhu Huihan
项目地点：武汉	Project Location: Wuhan
项目面积：1200 ㎡	Project Area: 1200 ㎡

本设计从建筑整体入手，将一个荒置多年的框架结构建筑重新整合。灰白、白、木质的深色便是整栋建筑全部的色彩，建筑的条形开窗如同中式画屏，从室内望去，一步一景，如犹抱琵琶半遮面般，极具美感。楼前开挖了水池，池水沿着池壁流动，余波荡漾，煞是生动，一派天人合一的景象。

整个门楼中庭的天顶完全打开，以中式天井为隐喻。天蓬的玻璃下使用了铜色铝板镂空树叶的纹饰，使阳光斑驳地洒入室内。中庭上悬挂的灯饰犹如荷花的苞蕾，又似一个个飘浮的孔明灯。建筑后部为保留一棵古树，特意规划出一处小天井，有效地利用楼梯将其围绕起来。古树冲出天井，呈现出一片生机，人树共生，恰到好处。

The design starts from the whole buildings and re – integrated a frame construction that abandoned for a long time. Gray, white, the dark color of the wood are the whole building's every color. The strip–type windows of the building look like Chinese painted screen, looking from the interior, it changes every turn. As if the beauty still half hidden her face behind a pipa lute, it is great beauty. In front of the building, water in the excavated pond flows along the pond wall and water rippling. It is truly vivid. How harmonious the scenery is.

The zenith of the gateway fully open and adopt the Chinese courtyard as a metaphor. Under the canopy glass, adopt coppery aluminium sheet hollow leaves as decoration, so make the dappled sunlight spilled into the room. Lights hanging on the Zhongting just like the lotus bud and like floating lanterns. In order to retain an ancient tree on the building rear, specially planned a small courtyard and effectively use the stairs circumfuse. The ancient tree grows out from the courtyard, so showing a life picture and we live with the tree, which be just perfect.

神韵 新中式空间设计典藏

高山流水的韵律之美
三明会所
The Beauty of Rhythm Water
Sanming Club

设计单位：福建品川装饰设计工程有限公司	Design Company: Fujian Pure Charm Decoration and Design Co., Ltd.
设计师：周华美	Designer: Zhou Huamei
项目地点：福建省三明市	Project Location: Sanming in Fujian Province
项目面积：1125 m²	Project Area: 1125 m²
主要材料：皮革、壁纸、大理石、铁刀木、银箔	Main Materials: Leather, Wallpaper, Marble, Indian Rose Chestnut, Silver Platinum Alloy

在这个融合了中式古典、现代简约与新古典风格的空间里，稳重、精致的色调及古典图案作为主轴线贯穿了整个空间。设计师用纯净、淡雅、明快的色调作背景，衬托出高品位的家具、灯具及艺术品陈设，结合突出的立体感和节奏感，烘托出会所高雅的文化氛围，这也成为贯穿会所各区域设计的基本准则。会所整体空间的装饰装修风格包容古今，设计师擅于在变化中寻求结合点，并努力贯彻到内部空间中，力求在追求整体统一的风格中体现出多元化、多视野的文化内涵。

基于这些高度统一的设计理念，会所的整体装饰呈现出豪华与高雅并重的装饰效果。此外，设计师在掌握全局装饰风格基调的基础上，又兼顾了开放、大众化的设计原则，同时又保持私密、个性的环境。使得人们无论是进入大堂，还是步入包厢，都能感知到其观赏性与艺术性并重的画面。

会所的整体空间采用大面积的大理石、壁纸及木质屏风隔栏，使得高贵优雅的气质呼之欲出，同时也让人欣赏到线与块面的流畅性。各种装饰元素的充分应用，让空间充满活力，使得大自然的气息也灵活穿梭其间。

In this space integrating classical Chinese, modern concise and neo-classical style, aesthetic classical elements were integrated into modern style. As the main axis, modest and exquisite color tones and classical diagrams run through the whole space. With pure, elegant and brisk color tone as the background, the design sets off high-taste furniture, lights and artistic layouts. Combined with highlighted sense of three-dimension and rhythms, the elegant cultural atmosphere of this club was displayed. And this becomes the basic principle for various spaces' designs of this club. The decoration of the whole club space incorporates ancient and modern style. The designer is good at searching for bonding points in these variations, and implements them into the interior space, thus to represent cultural connotations of diversification and multiple visions in attaining integral style.

Based on these unified design conceptions, the whole decoration of this club displays some decorative effects emphasizing on both luxury and elegance. Other than that, based on mastering well the whole decorative style, the designer pays attention to open and public style, while maintaining some private and personal environment. This makes it possible that people can perceive a picture of focusing on both outlook and artistic sphere while entering the lobby or the box rooms.

The whole space of the club applies large area of marble, wallpaper and wood screen partition, thus the noble and elegant temperament is vividly portrayed. At the same time, people can perceive the fluency of lines and surfaces. The sufficient application of various decorative elements makes the whole space appear more active and the atmosphere of nature is present in the space.

神韵 新中式空间设计典藏

东方情韵中的西式优雅
惠州高尔夫会所
Oriental charm of Western Elegance
Huizhou Golf Club

设计单位：KSL设计事务所	Design Company: KSL Design(HK) LTD.
参与设计：林冠成、温旭武、马诲泽	Associate Designers: Andy Lam, Wen Xuwu, Ma Huize
项目地点：广东省惠州市	Project Location: Huizhou in Guangdong Province
项目面积：10000 m²	Project Area: 10000 m²
主要材料：樟木、秀石、皇家木纹大理石、皮革、斑马木饰面、灰木纹大理石、黑钢、特殊玻璃、马赛克	Main Materials: Camphorwood, Imperial Wood Grain Marble, Leather, Zebra Wood Veneer, Grey Wood Grain Marble, Black Steel, Special Glass, Stone Mosaic

惠州高尔夫会所坐落于广东省惠州市惠东县大岭镇，地处高尔夫球场及湖边，风景秀丽，地理位置极佳。KSL设计事务所的设计力求让这个空间更加静谧而有深度，设计师对空间的功能分区进行了合理的布置，让客人无论休息或是用餐时，均能将高尔夫球场的美景尽收眼底。

项目的室内设计与建筑外观的中式风格相呼应，以新中式的折中手法，将具有国际感的现代家具及艺术装饰完美组合在一起，达到空间视觉上的统一。设计师注重质感的表达，希望在东方情韵浓厚的室内环境之中，也能够体现出西方人本主义的舒适和优雅，更是让客人在休闲舒适之外，能感受到内在的奢华与品位。质朴的材料和恢弘的空间，简约的装饰和精致的氛围，极致的对比给人以融汇中西、时空错落的非凡享受，更在山水相依的设计中，升华了人生的境界。

This project is located in Daling Town, Huidong County, Huizhou in Guangdong Province. This location along the golf course and lakeside is very picturesque and has wonderful geologic location. The design by KSL Design Firm tries to make this space much more serene and profound and makes proper arrangement towards the functional division of the space. Thus when the guest is having a rest or having meals, they can always enjoy the sceneries of the golf course.

The interior design of the project echoes the Chinese style of the architectural outlook. With new Chinese style approach, the designer perfectly integrates modern furniture with international feel and the artistic decorations, thus achieving the visual unity of the space. The designer emphasizes on the expression of the texture and hopes that the interior environment with intensive oriental charms can also display the comfort and elegance of western style. What is more, the customers can feel the inner luxury and taste from this leisure and comfort. The primitive materials and magnificent materials, concise decorations and delicate atmosphere and extreme contrast give people incomparable enjoyment integrating east and west, which is like in totally another world. Other than that, the design of mountain and water uplifts the realm of human world.

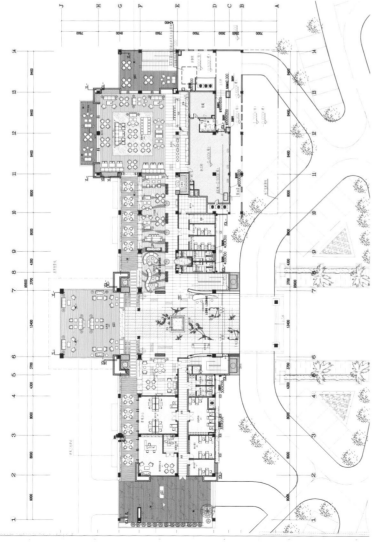

一层平面图　　　　　　　　　二层平面图

神韵 新中式空间设计典藏

思绪的留白
意兰亭休闲会所
Mind Blank
Yi Lanting Leisure Club

设计公司：中国(合肥)许建国建筑室内装饰设计有限公司	Design Company: China (He Fei) Xu Jianguo Architectural interior design Co., Ltd.
主设计师:许建国	Chief Designer: Xu Jianguo
参与设计：陈涛、欧阳坤、陈迎亚	Participants: Chen Tao, Ouyang kun, Chen Yingya
项目地址：合肥	Project Location: He Fei
项目面积：460 ㎡	Project Area: 460 m²
撰文：赵琳	Writer: Zhao Lin
摄影师:吴辉	Photographer: Wu Hui

门外，熙熙攘攘的人潮正渐渐退去。心里，一天的浮躁、烦躁也终于退去。当夜色来临，一切慢慢地归于平静的时候，情绪也由疲惫、失落慢慢地归于平静。独处的时候,总觉得丢了些什么，总希望能找到那些本该属于自己而恰恰失去的……

那砖，一层一层整齐地排列着，在墙与墙之间，连着那青石板巷道，一个烟雨朦胧的黄昏，一把小红伞……

那瓦，一片一片，似乎属于那个年代，似乎是那么的遥远，只是，一旦看到它，心里就自然涌起莫名的亲切和惆怅，似乎已经闻到了来自大地泥土的气息。

那木，是一条一条带着疤节的木条，看起来好像一点儿也不精巧，一点儿也不好看。不过，当想起了它的朴与拙，也就自然而然想起来那句话：君子与其练达，不如朴鲁。

Outside the door, people are walking away from this street. It is the same with the restlessness in your heart. When the day turns into night, one will calm down like all the other things. Along with the peace of night comes a sense of lost. A lonely person is always searching for something that should have belonged to him...

The stone lane, with walls composed of neat bricks on both sides, falls into the misty rain. There is someone under a small red umbrella wandering on it...

The tiles on the roof seem to have existed over years, which would arouse your sense of familiarity and melancholy towards the old age. While seeing it, you will feel your lung filled with a pleasant earthy smell.

The battens with wood knots seem to be too ordinary to attract your preference, but as a saying puts it, sometimes simplicity is better than sophistication.

神韵 新中式空间设计典藏

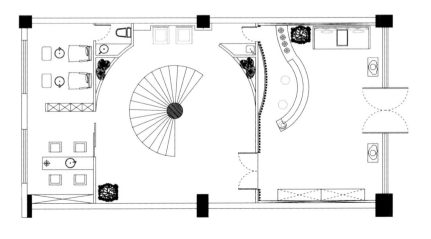

一层平面图

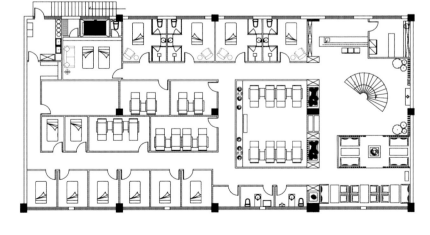

二层平面图

传统艺术的现代演绎
和谐苑餐饮会所
Modern Interpretation of Traditional Art
Harmony Court Restaurant Club

设计师：卓卫东	Designer: Zhuo Weidong
项目地点：福州	Project Location: Fuzhou
项目面积：2000 ㎡	Project Area: 2000 ㎡

本案是一座四层楼高的集餐饮与休闲为一体的会所。有别于以现代欧式为主的潮流设计，设计师抽取典型的中式风格的传统元素，将大量的传统符号进行标新立异的组合，在现代工艺、现代技术的演绎中塑造出一个东方空间。

设计师采用古典的大红、金黄为主色调，配以原木家具的色彩，契合了古代宫殿以红、黄为主的色彩美学，彰显出空间的华贵、恢宏之感。每个包厢设置了餐区与休闲区，休闲区域的布置焕发出令人神往的儒雅风韵。设计师在第二、三层的包厢区中设计了大量弧形的切面，使中规中矩的室内环境在曲形的线条、墙面甚至空间的点缀下不显死板。与之相对应的是蜿蜒回旋的过道，"幽"在这曲径之中，"美"在这回转之间，传承了古人"曲径通幽"的隐士价值。装饰材料中实木占据半壁江山，桌椅、案几、板凳等室内家具及吊顶、门窗等设备，在材质属性不变的基础上，通过肌理的处理、造型的多样化及颜色的细微差异让每一件单品各具特色。

This case is a four stories high club that set dining and leisure as whole. Different from the fashion design that based on modern European. Designer extracted typical Chinese style of traditional elements, combined a large number of traditional symbols unconventionally and has carved out a oriental space in the deduction of the modern technology and modern craftwork.

Designers adopted the classic red, golden as the predominant colors, log furniture color as the secondary color and corresponded to the color aesthetics of adopting the red and yellow as the dominant color of ancient palace, therefore highlighting the luxury of space and magnificent feeling. Each box was equipped with a dining area and leisure area and the layout of leisure area glows intriguing elegant charm. Designers designed a large curved section between the second and third boxes area, so that law-abiding indoor environment is not significant under rigid the decoration of curved lines, walls and even the space. Corresponding to it is winding whirly corridor that "quiet" in this meandering and "beauty" in the rotation, which heritage of the ancient "winding streets" the value of hermits. Hardwood decoration materials occupied half of decoration materials. Tables and chairs, end tables, wooden benches, and other indoor furniture and suspended ceiling, doors and windows and other equipment based on the same material properties made every single product has its own characteristics by the processing of texture, the diversification of the model and the slight difference of color.

神韵 新中式空间设计典藏

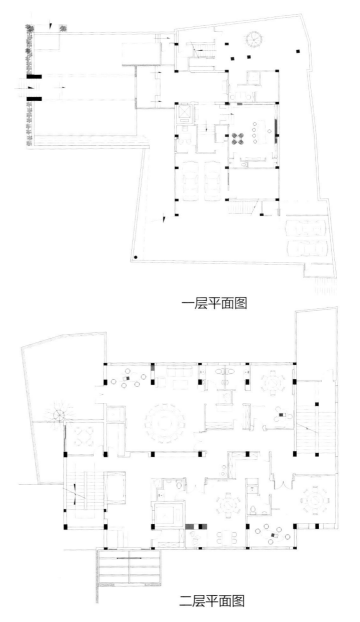

一层平面图

二层平面图

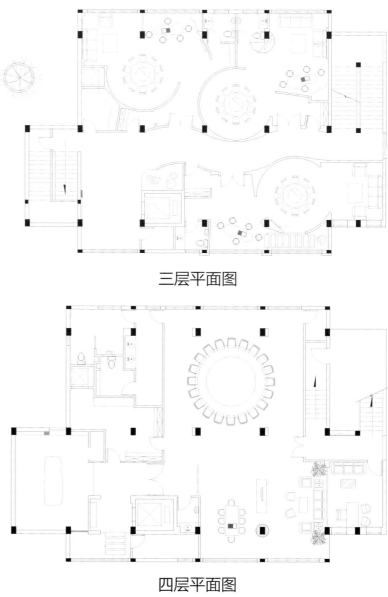

三层平面图

四层平面图

神韵 新中式空间设计典藏

后巷书生，感于斯文
琴南书院

Hou Xiang Scholar, the Feeling in the Gentle
Qin'nan Academy

设计师：吴奇
项目地点：福州
项目面积：300 ㎡

Designer: Wu Qi
Project Location: Fuzhou
Project Area: 300 ㎡

琴南书院的结构与普通的福州古民居没有太大区别，四周风火高墙，石框大门，门额上题着"琴南书院"四个大字。书院虽不大不深，但承袭了南方古宅的风格，进门依次是天井、回廊、左右厢房等。一进大厅是穿斗式木构架，进深五柱三开间，中为大厅，匾额上写着"畏天"二字，取的是林纾一生的座右铭——"畏天而循分"之意。庭院里，植数株花草点缀其间，淡雅清香之气，回旋而上。东西厢房中，古朴的书架，木制的茶台，窗明几净，灯光明暗适度。相视而坐，宜心自然。

The structure of Qin'nan Academy doesn't have much difference with Fuzhou ordinary ancient houses. Surrounded by wind-fire walls, stone door frame, the four characters — "Qin'nan Academy" are inscribed on the amount of doors. Though not deep, not large, the academy inherits the style of the south old house. Entering the door, there are patio, corridors, wing rooms around and so on. Entering the house, there are column-and-tie wooden construction, five columns and three bays. The hall is in the middle, with the words "Revere God" inscribed on the board, taking the meaning of Lin Shu motto—"Revere God, you will restrain yourself". The courtyard is dotted with a few plants and flowers, and elegant fragrance pervades in the air. In the wing room, the quaint bookshelf and wooden tea sets are bright and clean. The light is moderate. Sit face to face, should be a beautiful picture.

The Second Chapter
Residential Space

第二章
住宅空间

亦古亦今 满室芳华
中星红庐65#别墅
Ancient and Modern Full Chamber Youth
Star Red House 65 # Villa

设计公司：上海鼎族室内装饰设计	Design Company: Shanghai Ding Family Interior Design
设计师：吴军宏	Designer: Wu Junhong
项目地点：上海	Project Location: Shanghai
摄影：三像摄 张静	Photography : Three Photo Photography Zhang Jing

推门而入，玄关背景墙后的公共区域以两列纵向排开的圆柱分隔，小客厅与餐厅则通过半开放式的栅栏遥遥相对，三角梁样式的吊顶让人感觉仿佛置身于老式住宅的建筑中。同一屋檐下的大家庭氛围，使得业主一家无论是饮茶还是用餐，都备感温馨。

整个空间的设计，均为古色古香的中式风格，而在格局和功能上，则以现代生活为蓝本，突出度假和休闲的主题。玄关左侧的区域是主客厅，再往里深入，便是摆放着文房四宝的书房。玄关的右侧，是连接地下室至二楼的旋转楼梯。考虑到老人行动不便，设计师贴心地把老人套房安置在一楼。二楼是包括家庭室、主卧室及两件客房的私人空间，地下室则是丰富的公共活动区，有收藏室、茶室、影音厅，还有SPA室等休闲区域。

穿行室内，仿若行走在曲折、幽邃的小径，别有意境。首层的走廊间，在大理石地面及深色镜面天花、墙面的默契配合下，整条走廊深远而悠长。地下室的茶室内，镜面顶棚则将展示柜内的陈列品无限延展，既扩大了室内空间，也丰富了空间语言。设计师在格局和功能上以现代生活为蓝本，打造出一个亦古亦今的休闲生活空间。

Push the door, the public areas after the hallway wall are separated by two rows open cylindrical. small living room and dining face each other by semi-open fence. triangle beams style ceiling make people feel like being in the old building. Family atmosphere under the same roof makes the owners have a sense of warmth either tea or a meal.

The whole design of the space is antique Chinese style.The structure and function are based on modern life, highlighting the holiday and leisure theme. Area to the left of the hallway is the main living room, go down to depth, there is stocked with four treasures of the study. The right of the hallway is spiral staircase to connect the basement and the second floor. Taking into account the mobility of the elderly, the designer thoughtfully places the elderly's suite on the first floor. The second floor is a private space including family room, master bedroom and two guest rooms. the basement is a wealth of public activity area-collection room, tea room, video room, as well as SPA room and other leisure areas.

Walk through the room, like walking in the twists and turns, deep and quiet trails. Between the hallway on the first floor, under the tacit cooperation of marble floor and dark mirror smallpox, metope, the entire corridor is far-reaching and long. In the tearoom of basement, mirror ceiling will display unlimited extension of the cabinet in the showcase, both to expand the interior space, but also enriches the language of space. On the structure and function, designer models modern life and creates a both ancient and modern leisure living space finally.

神韵 新中式空间设计典藏

神韵 新中式空间设计典藏

神韵 新中式空间设计典藏

神韵 新中式空间设计典藏

蓝调贵族
融侨新城泷郡别墅样板房
Blues Aristocracy
Rongqiao Legend Villa Show Flat

设计公司：上合设计顾问有限公司	Design Company: Shanghe Design Consultants Co., Ltd.
设 计 师：余周霖	Designer: Yu Zhoulin
参与设计：叶志应、王舟	Associate Designers: Ye Zhiying, Wang Zhou
项目地点：福清	Project Location: Fuqing
项目面积：850 ㎡	Project Area: 850 ㎡
主要材料：树瘤木烤漆、皮革、镜面、不锈钢、大理石、实木复合地板	Main Materials: Burl Wood Paint, Leather, Mirror Surface Stainless Steel, Marble, Solid Wood Composite Floor
摄 影：三像摄 张静	Photography: Sanxiang Photography Zhang Jing

独栋别墅结构的户型应该是生活空间最为宽裕、舒适的户型结构，这样的空间就不仅仅是满足吃饭、睡觉等简单的家庭活动，多样化的功能分区丰富了人们的家居生活。本案空间还配有电梯，方便主人在家中上下活动，进入家中就等于进入了一个私家的小世界。

一层作为公共空间，是主人招待客人的地方。客厅与餐厅比邻，客厅的挑高高达2层楼，装上华丽硕大的水晶吊灯，也丝毫不觉突兀。整面的落地窗子，让空间显得格外透亮，而窗外的美景也自然地成为背景。客厅的基调十分简约，围绕一圈的沙发是客厅的主体，水蓝的绒布沙发，展现出最优雅的气息。石料是空间的主材，用各式模样、颜色的大理石来装点空间，沉稳的氛围使空间大气而美观。

本案配有多间卧室，每间卧室都是一间设备齐全的小型套房，配备有休憩区、更衣间、大型卫生间等，完善的配备让主人的生活更加便利，生活更加舒适。

另外，酒吧、茶室、娱乐室一应俱全，邀请一群好友小聚也完全可以满足。整个居室空间以利落、时尚的现代风格演绎了格调家居的气质，使居住至上的理念贯穿始终。

The house type for single villa should have the most comfortable and most cozy structure of life space, which shall not just meet with simple family activities such as having meals and sleeping. The multiple functional space divisions enrich people's home residential life. This space has elevators, convenient for the host to move upwards and downwards inside home. Entering home is like stepping into a private little world.

As a public space, the first floor is a place for the host to receive guests. The living room neighbors the dining hall. The living room is as tall as 2 floors. The space is decorated with grand and magnificent crystal drop-lights, which make the space do not appear abrupt at all. The whole French window makes the space appear quite bright and the scenery outside the window naturally becomes the background. Stone is the main material for the space. The space is decorated with various formats and various colors of marble. The sedate atmosphere makes the space appear grand and magnificent.

This project is attached with many bedrooms, each bedroom is a tiny suit of complete facilities, collocated with leisure space, changing room and grand washroom. The complete equipment makes the master's life become more convenient and life become more comfortable.

Other than that, the space has wine bar, tea room and entertainment space. It could make do for a group of friends to gather here. With brisk and fashionable modern style, the whole residential space displays the temperament of style residence, making the concept of "residing first" go through the whole space.

神韵 新中式空间设计典藏

神韵 新中式空间设计典藏

神韵 新中式空间设计典藏

神韵 新中式空间设计典藏

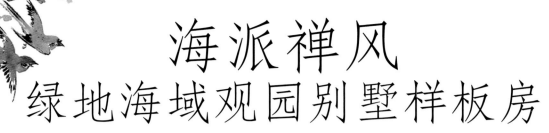

海派禅风
绿地海域观园别墅样板房
Shanghai Zen View
Green Sea View Garden Villa Show Flat

设计公司：上海风锐设计工程有限公司	Design Company: Shanghai Fengrui Design Engineering Co., Ltd.
设计师：胡斐	Designer: Hu Fei
设计助理：蔡翔	Design Assistant : Cai Xiang
项目地点：上海嘉定	Project Location: Shanghai Jiading
项目面积：350 ㎡	Project Area: 350 ㎡
主要材料：雅士白大理石、意大利灰大理石、银白龙大理石、镜面黑色不锈钢、麻草壁纸、真丝壁纸、影木、黑檀木、真丝布艺等	Main Materials: Aston white marble, Italian gray marble, white marble dragon, mirror black stainless steel, hemp grass wallpaper, silk wallpaper, shadow wood, ebony, silk cloth, etc.
摄影：三像摄 张静	Photography : Three Photo Photography Zhang Jing

绿地海域观园是绿地集团倾力打造的高端别墅区，有六合院别墅、联排别墅等，是近20年来绿地集团为嘉定新城打造的原创海派新中式别墅群。临水而居的生态环境，创新的中式建筑及新中式园林都具有独特的魅力。

室内设计师将简约与传统的中式元素重组，目的是塑造一个外在形式现代时尚、内在气质内敛东方的新中式高端住宅样本。传统的中式风格偏向深色调，且加入过多的传统家具及饰品，容易产生沉闷、守旧的感官体验。本案设计希望另辟蹊径，尝试以白色调表达中式时尚清新的一面。设计师以白色大理石做底，黑色金属线条勾边，用类似工笔画线框白描的手法表达天、地、墙的空间关系。局部重点区域用拼花灰色大理石划分。客厅顶部用白色小方格造型表达东方意境，配以同样形态的现代吊灯，简练的造型和色彩使整个客、餐厅呈现对比鲜明的黑白色调，高雅大气。

在客厅的主要端景墙面，为了调和偏冷的色调，设计师采用了少量明黄色高光木皮来增强温暖的气氛，餐厅背景装饰柜也用了特别的银白色闪光贝壳饰面，以增强奢华的灰色质感。开放式的吧台，在造型上也隐约有明代案几的影子，整体造型手法简练，比例工整，色调材质层次分明，张弛有度。

三楼的主卧是另一大亮点，透明的采光天窗堪比东南亚海景度假酒店房，大面积的青砖墙面结合外墙的银灰洞石，与水面的青石交相辉映，阳光穿过大面积的格栅照进浴缸，一个亲近自然、光影灵动的SPA空间足以打动追求高品质生活的现代都会女主人。

这套别墅的软装配饰也充分体现了设计师要求的外在时尚、内在东方的精神内核，大到吊灯、家具，小到针线、纸笔，无不和硬装相得益彰。

Greenland Sea View Park is high-end villa build by Greenland Group with great efforts. there are Kuni countryard villas, townhouses, which is the original Shanghai new Chinese villas created by the Greenland Group for Jiading in the nearly 20 years. Ecological environment with waterfront neighbors, innovative Chinese architecture and the new Chinese garden all have unique charm.

The interior designer reorganizes the brief and traditional Chinese elements, aim to create an Eastern new Chinese high-end residential sample with external form of modern fashion and the inherent qualities introverted. The design of this case hope to find another way, trying to express Chinese fashion fresh using white tune. Designers do at the end of white marble, black metal lines Crochet, expression of heaven, earth, Wall Painting spatial relationships in a similar way wireframe line drawing. Small white box modelings in the top of the living room express Orient mood, with modern chandeliers in the same form, concise style and color, which makes the entire dining room present a clear black and white, elegant and atmosphere.

In order to reconcile the cool colors, the main side view wall in the living room, designers use a small amount of bright yellow high-gloss veneer to enhance the warm atmosphere. The shape of open bar is also looming shadow of Anji of Ming Dynasty. The overall shape is in a concise way, the proportion of neat, distinct material tone, elastic.

The master bedroom in the third floor is another highlight. Transparent skylight is comparable to Southeast Asia sea-view resort room. A large area of brick walls combined with local facades silver travertine. Sunlight shines into the bathtub through large areas of grill. A close to nature, blight SPA space is sufficient to impress the modern metropolis hostess who pursuits high life quality.

The soft assembly of the villa is also fully embodies the external fashion designer required and inherent Oriental spiritual core. From chandeliers and furniture to needle, pen and paper all bring out the best in each other with hard assembly.

神韵 新中式空间设计典藏

神韵 新中式空间设计典藏

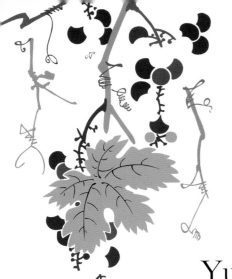

岭南气质
保利越秀·岭南林语F2别墅
Lingnan Temperament
Yuexiu Poly · South of Lingnan Lin language Villa F2

设计单位：广州市韦格斯杨设计有限公司	Design Company: GrandGhostCanyon Designers Design Co. Ltd.
项目地点：广州开发区	Project Location: Guangzhou Development Zone
项目面积：304㎡	Project Area: 304㎡
主要材料：郁金香大理石、哥伦比亚啡大理石、清水玉大理石、砖、橡木饰面	Main Materials: tulip, Columbia brown marble, jade marble, brick, water oak veneer

本案别墅沿山而建，可以零距离亲近自然山体。本户型为三层结构，设计中利用架空层和夹层的面积，使其形成一个舒适的四层别墅。

在设计风格上采用现代中式的设计手法，融入简练的中式元素，将含蓄内敛与随意自然两种气质完美结合起来，整体基调以舒适、轻快为主，既符合人们的使用要求，又能充分展现中式的韵味。

负一层作为生活功能的完善空间，针对居住者是古董爱好收藏者、喜欢茶饮的特点，在设计上体现出浓列的文化气息，体现主人沉稳，喜欢中国传统文化的性格特点，软装上体现空间中"禅"的韵味。

首层为会客、就餐区域，设计师充分利用原有的建筑形态，装饰细节上崇尚自然情趣，将花鸟虫鱼与石材等元素相结合，在现代装饰手法的诠释下，体现出中国传统的美学精神。

二层为卧室，设置长辈房和儿童房，体现一家人其乐融融的氛围。

三层为主人房，作为主人的私密空间，舒适性是关键的要件之一，通过简练的线条与软包相结合，展现一个内敛而舒适的睡眠空间，再搭配上具有中式文化特征的元素，体现居住者的性格特征，展现出另一层次的生活环境。阳光花房，让使用者的空间从室内延伸到室外，为女主人提供了一个独立的活动场所。

神韵 新中式空间设计典藏

神韵 新中式空间设计典藏

In this case the villa built along the mountain, you can zero distance close to natural mountain. This apartment layout is divided into three layers, layer overhead and dissection of the area by use of the design, which forms a comfortable four storey villa.

By the design of modern Chinese style in the design style, Chinese style elements into concise, the implicative combined with the random nature of two temperament perfect, the overall tone of comfort, light to give priority to, not only to meet the requirements for people to use, and can fully demonstrate the charm of chinese.

A negative life as a function to perfect space, according to the occupant is an antique loving collectors, like tea music characteristic, manifests the cultural atmosphere thick columns in the design, reflect master calm, like Chinese character of traditional culture, embody the space soft outfit "Zen" charm.

The first floor is the reception the dining area, make full use of the original architectural form, decorative details on advocating natural taste, flower and bird painting and stone combination of such elements, in the modern techniques of interpretation, reflect the aesthetic spirit of traditional Chinese.

The two layer is the bedroom, set the elders room and children's room, embody one family enjoyable atmosphere.

The three layer is the owner of the housing, as the owner of a private space, comfort is one of the key elements, through the combination of lines and soft concise phase, showing a restrained and comfortable sleeping space, soft outfit collocation with Chinese cultural elements, embodying the characteristics of occupants, show another level of life environment. Sunshine, let the user space extend from indoors to outdoors, provides an independent activity space for the hostess.

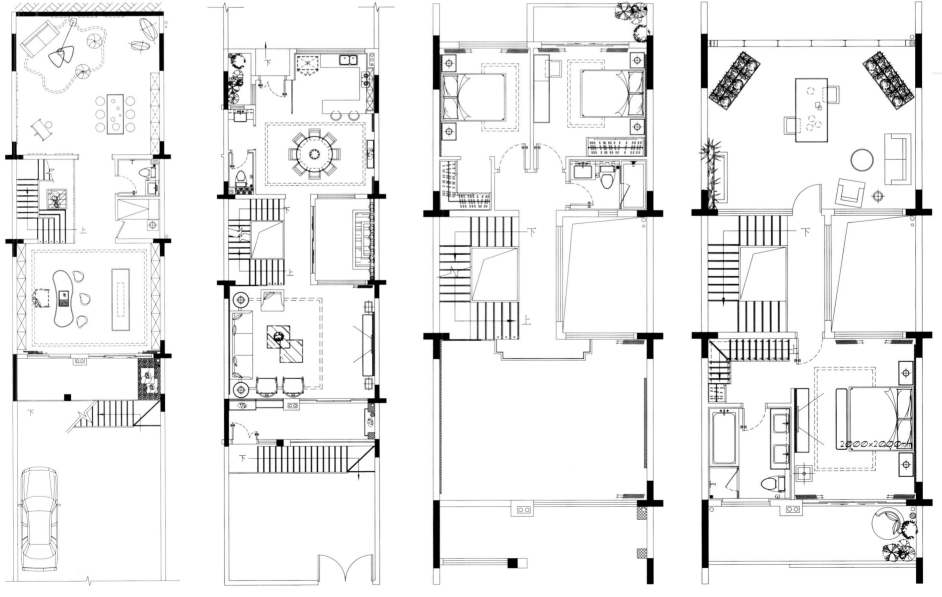

负一层平面布置图　　一层平面布置图　　二层平面布置图　　三层平面布置图

神韵 新中式空间设计典藏

原木清风
赣州中洋公园首府样板房设计
Log and Breeze
Ganzhou Zhongyang Park Capital Show Flat Design

设计团队：KSL设计事务所	Design Company: KSL DESIGN(HK)LTD.
主设计师：林冠成	Main Designer: Andy Lam
项目面积：145 ㎡	Project Area: 145 ㎡
项目地点：江西赣州	Project Location: Jiangxi Ganzhou
主要材料：金贝壳、水曲柳、黑檀木纹、皮革、壁纸板、古铜	Main Materials: Gold Shell, Fraxinus Mandshurica, Ebony Wood, Leather, Wallpaper, Copper Plate

取自天然的设计素材最具有传统中式的神韵，如琢如磨的细腻工艺让整个样板房的设计显得质朴而清雅，原木材料的大量运用带来如沐清风的舒展惬意感。少许禅意的营造提升了空间的文化韵味，一丝不苟的线条让空间大气而通透，色调把握与灯光照明极为协调，恰到好处。

玄关：一把矮椅，一杯香茗，独特的意境随之而来。客厅：以浅色木质材料主打的色彩搭配提高了空间的亮度，花鸟壁纸与木纹屏风交相辉映，古瓶瓷器与仿清家具相映成趣，原木的天然清新营造出的空间的清丽而雅致。

餐厅：明清餐桌椅造型简洁，藏青色餐具与灿烂黄花的搭配让就餐环境极为优雅。餐厅与客厅之间用仿古长桌搭配古铜饰品进行分割，过渡自然且不失精巧。

书房：整个墙面设计成博古架，曲折的线条具有传统书房的墨香古韵气息，书架后的镜面装饰则让空间更显宽敞。

卧室：保持与客厅一致的格调，天然木纹饰面与麻灰壁纸的色彩相协调，浅灰的床品装饰质感极佳，飘窗的改造让整个空间精致而舒心。

From the design of material nature's most traditional Chinese charm, as a distinguished fine grinding process for the whole model of the design is simple and elegant, the use of a large number of log materials such as Mu breeze brings the comfortable stretch. To create a little Zen promotion space cultural flavor, be strict in one's demands line lets space atmosphere fully, grasp of color and lighting is very harmonious, be just perfect.

Inside: a low chair, a cup of tea, unique artistic conception attendant.Living room: in a light coloured wood materials main color matching improves the spatial brightness, flower wallpaper and wood screen add radiance and beauty to each other, the ancient porcelain bottle and Qing furniture gain by contrast, log natural fresh elegant Qingli create space.

Restaurant: the Ming and Qing Dynasties mensal chair shape concise, navy blue and bright yellow tableware mix make extremely elegant dining environment. Between the restaurant and living room table with antique bronze ornaments collocation segmentation, natural transition and without losing the delicate.

Study design: the whole wall into the shelf, zigzag lines quite traditional study ink rhyme, the bookshelf of decorative mirror lets a space more spacious.

Bedroom: keep consistent with the living room style, natural wood finishes and hemp gray wallpaper color coordination, grayish bedding decorative excellent texture, Piaochuang transformation so that the whole space is delicate and comfortable.

神韵 新中式空间设计典藏

平面布置图

神韵 新中式空间设计典藏

清雅灵韵 内敛悠长
深圳·红树湾现代风格别墅样板房
Elegant Aura,Introverted Long
Shenzhen•Mangrove Bay Modern Style Villa Show Flat

设计团队：KSL设计事务所	Design Company: KSL DESIGN(HK)LTD.
主设计师：林冠成	Main Designer: Andy Lam
项目地点：广东深圳	Project Location: Shenzhen Guangdong
项目面积：775㎡	Project Area: 775㎡
主要材料：橡木饰面、橡木实木地板、白色乳胶漆、夹丝玻璃、皮革、编织墙纸、进口灯具、进口地毯	Main Materials: oak veneer, oak wood floors, white latex paint, wired glass, leather, woven carpet, wallpaper, imported lamps imported

大象无形，不事雕琢，灵韵悠长。此别墅样板房的设计去除了繁复的装饰，雅白与橡木饰面的天花墙面搭配大理石的色泽纹饰，清雅的自然韵味内蕴深长。挑高的空间设计及灵活巧妙的布局，塑造出大宅的稳健气度。极富质感的进口陈设架构出空间的人文气质，使之焕发出空间的独隽神采，晕染出深隽内敛的现代简约空间。

客厅：挑高的空间设计使空间显得明亮大气，在柔和灯光的搭配下，精美奢华的进口灯具与皮革等高档材质相互辉映。

书房：以木色为主、灰色为辅的基调营造了古朴安静的书香氛围，清新的蓝色为视觉增添了另一番惊喜，简洁而有力度的设计提升了空间的格调。

二层卧室1：天花和橡木实木地板打造出简单舒适的休憩空间，湖蓝石墩和白陶瓷碗在灯光下越发剔透，整个空间看似不经意的搭配，却清醇而又悠远。二层卧室2：线条简练质朴的实木家具和精致的陈设在灯光的烘托下更显气韵，无阻碍的日光窗景从落地窗向室内延伸，让身心享受微风轻拂的自在和惬意感。

三层主卧：独特的挑高空间以及三角顶棚造型，巧妙地带来了截然不同的睡眠享受。整面的落地窗营造出通透敞亮的生活空间，光影的流动使空间显得更加静谧。

浴室：简洁流畅的设计线条勾勒出浴室的整洁与自在轻松，让繁忙一天的身心得到舒缓。

负一层红酒雪茄吧：博古架和奢华皮革家具带来文化的多重韵味，璀璨剔透的水晶灯让古典韵味与现代简约巧妙兼容。

The great form has no shape., the matter does not carve, aura long. The model room villa design case of removal of complicated embellishments, elegant white and oak veneer smallpox metope is tie-in marble color decoration, elegant natural charm of intrinsic deep. Pick the space design of high and flexible layout clever, shaped mansion of robust tolerance. The very texture of imported furnishings architecture space refinement and humanistic temperament, glows the multiple cultural implication of expression, blooming out deep Jun restrained modern minimalist space.

Living room: pick the space design of meal of high explicit bright atmosphere, in the soft light tuning, exquisite luxury imported lamps and leather and other high-grade material has embraced.

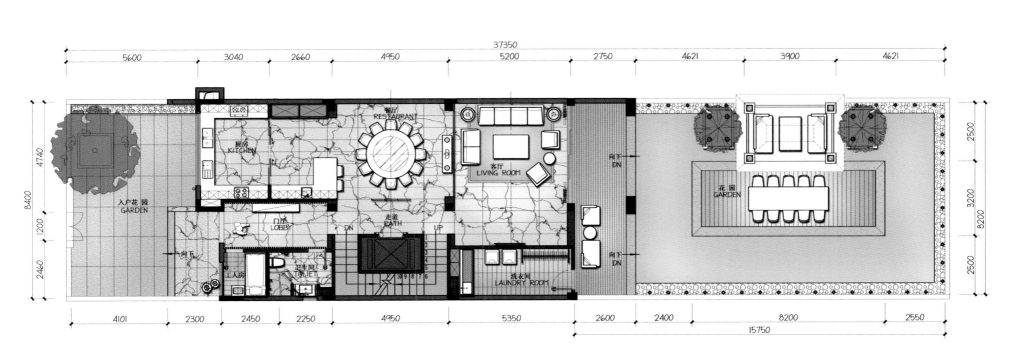

一层平面图

二层平面图   三层平面图

Study: wood color based, supplemented by creating simple grey tone quiet elegant atmosphere, fresh blue give visual another surprise, simple design and efforts to improve the space style.

The two floor bedroom 1: Practice of smallpox and oak wood floors to create a simple and comfortable rest space, blue stone and white ceramic bowl in the light more condensate extraction clear, whole space appears to match the casual, but mellow and distant.The two floor bedroom 2: concise lines and plain solid wood furniture and elegant furnishings in the lamplight foil under the more matter rhyme, sunlight window view unobstructed extends to the interior from the window, let the body and mind to enjoy a comfortable and cozy feeling cool and breezy.

The three floor master bedroom: unique high open space and stable triangle smallpox modelling, subtly to different sleep enjoy. Floor to ceiling windows create a transparent light and spacious living space, light flowing along with time more quiet.

Bathroom: design line succinct smooth outline of the bathroom clean, comfortable, let a busy day and get the precipitation diastolic pressure.

Negative a layer of red wine & Cigar Bar: A. Keita bocoum aircraft and luxury leather furniture brings multiple flavor of culture, crystal lamp bright clear that classical charm and modern minimalist ingenious and soft.

静语凝思
上海长城珑湾样板房
Static Language Meditation
Shanghai Great Wall of Long Bay Open Model Houses

设计公司：KSL设计事务所	Design Company: KSL DESIGN(HK)LTD.
设计师：林冠成	Designer: Andy Lam
项目地点：上海	Project Location: Shanghai
项目面积：170㎡	Project Area: 170㎡
主要材料：灰橡木、灰茶镜、黑钢、麻质壁纸、白砂石、木地板、白洞石、实木屏风	Main Materials: gray oak, ash tea mirror, black steel, linen wallpaper, white sand, wood floors, white travertine, wood screens

本案设计构思严谨，用色稳健大气，色彩上以代表稳重、成熟的马鞍棕色为底，配以理智、安详的藏蓝，彰显出高档的空间格调。空间上运用灯光和墙面造型把控室内气氛，营造出静谧、祥和的居家环境，使人沉浸其中。

客厅：以棕色、黑茶、藏蓝为主打色彩，营造出安静沉稳的空间氛围。空间自上而下都以中式风格为主题，顶棚使用复古吊灯，墙壁覆以古色古香的花鸟壁纸，又以黑色木格屏风来划分空间，使整个客厅充满浓郁的古典气息，内蕴深长。

餐厅：使用天然的木质地板与麻质壁纸，使就餐环境充满自然清新的味道，一侧的墙壁以灰茶镜装饰，不仅规避了餐厅较为狭长的不足，而且迎合了整体色调，形成视觉上的放大效果，使空间更为宽敞。

书房：推开木格屏风，一间简洁的书房呈现在眼前，规整的家具与饰品体现厨主人严谨的生活态度，墨绿色块装饰的书架让色彩产生跳跃感，中和了严肃的气息。

卧室：卧室以舒适安静为设计主题，色调上延续整体的棕色，以浅棕和米色烘托氛围，黑色木格配以造型典雅的灯具装饰床头，点明了中式风格的主题。

儿童房：浅草绿的窗帘和装饰画活跃了空间，玩具柜的悬空设计极为巧妙，柜底中一群探头探脑的小象呼之欲出，带给人无限惊喜。

The conception of this case is rigorous. Coloring is steady and atmospheric. Sedate, mature saddle brown as the grounding, together with the rational, peaceful dark blue highlights the high-end space style. The use of lighting and wall modeling controls the indoor atmosphere, creating a home environment and letting people immerse in it.

Living room: give priority to brown, black tea, dark blue color, creating a quiet and calm atmosphere. From top to bottom, space is in Chinese style, ceiling using vintage chandelier, walls covered with antique bird and flower wallpaper, black wooden lattice screens dividing the space, which makes the whole living room full of rich classical flavor, and the implication is profound.

Restaurant: The using of natural wood floors and linen wallpaper makes dining environment full of natural fresh taste. One side wall is decorated with gray tea mirror. It not only avoid the restaurant relatively narrow, but also meet the overall tone and form a magnified visual effect, making space more roomy.

Study: Open the wood lattice screens, a simple study shows. Structured furniture and accessories reflect the owner rigorous attitude to life. Dark green block decorated shelves product a sense of jumping, and neutralize solemn atmosphere.

Bedrooms: Comfortable and quiet is the design theme of bedroom. The tone continues the overall brown. Light brown and beige foil the atmosphere. Black wood frame with elegant lamps decorating bedside dots the Chinese style theme.

Children room : Light green curtains and decorative painting enliven the space. Floating design of toy cabinet is very clever. a small group of poked around elephants are vividly portrayed in the bottom of the tank, and give people infinite surprises.

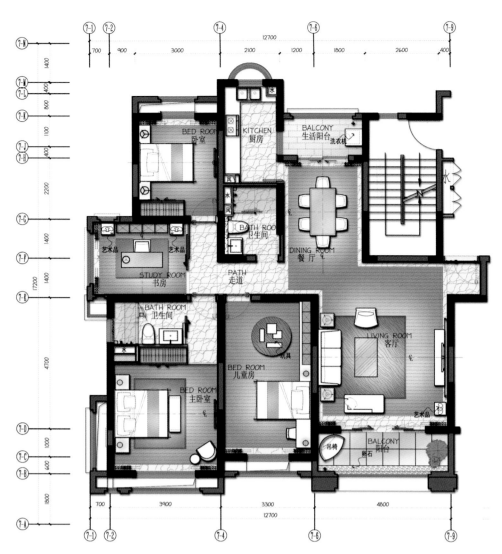

平面布置图

神韵 新中式空间设计典藏

神韵 新中式空间设计典藏

梅花三弄
央筑花园洋房样板房设计
Plum-blossom in Three Movements
Central Building Garden Villa Show Flat Design

设计团队：KSL设计事务所	Design Company: KSL DESIGN(HK)LTD.
设计师：林冠成	Designer: Andy Lam
项目面积：210 ㎡	Project Area: 210 ㎡
主要材料：橡木、古铜、黑檀木地板	Main Materials: Oak, bronze, ebony floor

中式风格的古朴韵味和典雅气度总令人心神往之，但传统中式的古旧和沉闷却常令现代都市人难以驾驭。本案使用了丰富的经典中式元素，将之改良形成全新的现代中式风格。大量橡木材质的使用营造出自然舒展的格调，整体空间设计手法张弛有度。

客厅：天然木材大量出现在客厅的设计中，木料本身的色彩和纹饰营造出温良敦厚的居家氛围。黑檀木地板的深沉色彩配以顶棚的明亮色调，整体自然而和谐。从空间布局到家具造型，硬朗的线条都形成了规整的空间感，墙壁的装饰画则提升了空间的意境。

餐厅：简洁造型的餐桌选用了深咖色系，一束蓝色的鲜花装点出浪漫的空间氛围。

楼梯：楼梯设计最忌暗淡，浅色的地板和璀璨的灯饰瞬间提亮了整个空间，通体木制条纹装饰的墙壁则在形成完美分区的同时，带来了淡雅的自然气息，在灯光的配合下显得明亮而清新。

儿童房：墨绿和条纹是儿童房的装饰元素，从窗帘色调到床品选择均一丝不苟，清新中带着俏皮，宁静中带着温馨。

卧室：卧室的空间氛围更突显温馨感，木纹贴面的墙壁让人心神安宁，或横或竖的几何线条不时出现在顶棚、墙壁和家具上，穿越时空的古韵感扑面而来，舒展而温暖。

The Chinese style of the ancient charm and elegant tolerance always makes people fascinated, but modern urban people hard to handle the old and dull in traditional Chinese style. The case adopted rich presents classical Chinese elements and ameliorated to form the modern Chinese new style. Massive oak materials used to create the natural and comfortable style and whole space was designed skillful tact and a degree of relaxation.

Living room: There are a lot of natural wood materials in the living room design materials and colors and patterns of wood itself to create a wonderful home atmosphere. Ebony floor deep color matched with the bright tone of ceiling makes the whole of space nature and harmony. From the spatial layout to the furniture modeling, hale line have formed a regular space, decorative painting of the wall enhance the artistic conception of space.

Dining room: Concise form table was chosen dark brown color and a bouquet of blue flowers decoration created romantic space atmosphere.

Stairs: Staircase design most abstain from dim. Light wood floors and bright light and lamps instantly brightens the whole space. The whole stripe decorating the walls formed a perfect partition, at the same time, brought the natural breath, matched under the light looks like bright and fresh.

Bedroom: Space atmosphere of bedroom manifests more warm and comfortable. Wood walls makes people the peace of mind, horizontal or vertical geometric lines appeared from time to time in the ceiling, walls and furniture, through time and space of the feeling emerged, comfortable and warm.

Kids room: Dark green and stripes are the active elements of kids room. From the choice of curtains color to bed goods are conscientious and meticulous. There are fresh with playful and peace with warm.

神韵 新中式空间设计典藏

神韵 新中式空间设计典藏

神韵 新中式空间设计典藏

泼墨与白描的新诠释
中航复式A2-1

The New Interpretation of Ink and Line Drawing
CATIC Duplex Apartment A2-1

设计单位：郑树芬设计事务所	Design Company: Simon Chong Design Consultants Ltd.
主创设计师：郑树芬	Main Designer: Simon Chong
软装设计师：郑树芬、杜恒、杨立	Soft Decoration Designers: Simon Chong, Amy Du, Yang Li
项目地点：贵阳	Project Location: Guiyang
项目面积：230 ㎡	Project Area: 230 ㎡

我们意在通过新装饰主义与中式传统经典元素的巧妙兼容，去感受国际文化交织的力量，强调一种超越世俗的生活境界，让生活在稳重、含蓄、内敛中自由绽放……

此户型为复式楼，一层有较大的入户花园，穿过入户花园就是客厅和餐厅。餐厅的一面墙采用了木架窗花造型作为装饰，不少家具采用了实木，但沙发却采用布艺，意在给空间营造轻松舒服的氛围，这也许就是设计想要表达的现代中式的韵味。值得一提的还有客厅的地毯采用了烟灰色，不仅与窗帘的色彩相呼应，而且极具品质感。单人沙发的布艺采用了回形纹的中式元素，使空间氛围呈现出的不仅是中式韵味，而且极具时尚奢华的东方感，同时又传达出安静优雅的尊贵氛围。

主卧以灰白色为主基调，床头的艺术画框极其生动，鸟儿的停留似乎让生活静止下来，加上简单陈设的艺术布置，给人极强的东方韵味及舒适感。儿童房的色彩给人以天真活泼的感觉，床头的油画将一种简单的童真又提升到了另外一个境界。

不管是一层还是夹层的书房，整个空间就像一张洁白的宣纸，设计师在上边渲染着豪放的泼墨和纤细的白描，将中国文化的万千情怀与东方人的审美情趣完美地结合在一起。无论哪个空间，灰白、木色成为空间的主色调，视觉的搭配将当代与古典巧妙地结合，客厅茶几上的中式茶具、陶瓷瓶、中式餐厅吊灯表达了设计师对汉文化的热爱。而时尚的家具、水晶灯又为空间增添了不少现代感。设计师对多种物料的尝试让空间有了不一样的视觉感受，极大地丰富了空间的视觉感受。

We aimed to clever compatible through decoration of the new Chinese classical elements, to experience the culture interweaveenergy, emphasize a kind of beyond the realm of life the secular, let life shine free in the stable, reserved, introverted……

This apartment layout for the penthouse floor, a layer of a garden largerhouseholds, through the garden is the restaurant, the restaurant wall used wooden grilles modeling as a decoration, a lot of furniture using wood asmaterial, but it uses the fabric sofa, intended to give space comfortableatmosphere, perhaps this is the desired expression the modern Chinesecharm. Also worth mentioning is the living room carpet, the gray smoke, not only with the curtain color echoes, and a very sense of quality. Sofa Chinese elements to form grain, make the space atmosphere showsnot only Chinese flavor, but also very stylish luxurious Oriental feeling, alsodisplay the quiet and elegant noble ambience.

Master bedroom with gray tone, the art frame extremely vivid, the birds stayseemed to make life down to rest, with simple furnishings art layout, givepeople a strong oriental flavor and comfort.The children of color to the innocent and lively feeling, the painting will be a simple child and ascend into another realm.

No matter what space, gray, wood color become the main space,visual mix of contemporary and classic combination, the living room side table, Chinese tea set, ceramic bottle, Chinese restaurant chandelierexpress designers love of Chinese culture. But stylish furniture, crystal lampfor the space added a lot of modern. Try on a variety of materials in space with different visual feelings, greatly enriched the spatial visual experience.

神韵 新中式空间设计典藏

神韵 新中式空间设计典藏

神韵 新中式空间设计典藏

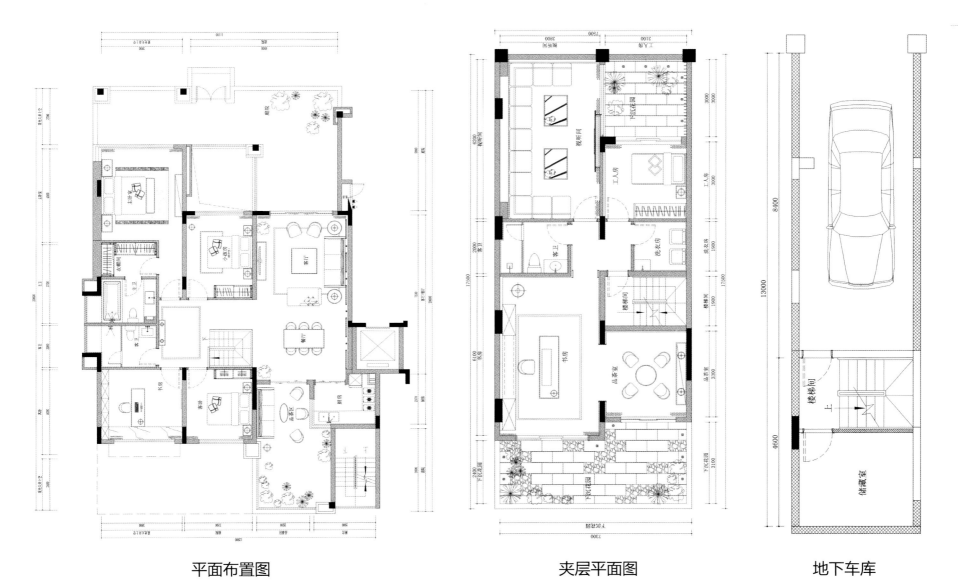

平面布置图　　　　　　　　　夹层平面图　　　　　　　　　地下车库

如诗梦，似古今
中央公园二期11-B01

Such as Poem and Dream, Like Ancient and Modern
Central Park, Phase 2, 11-B01

设计公司：KSL设计事务所	Design Company: KSL DESIGN(HK) LTD.
设计师：林冠成	Designer: Andy Lam
项目地点：广东深圳	Project Location: Shenzhen of Guangdong Province
项目面积：148㎡	Project Area: 148 m²
主要材料：白砂石、橡木、黑檀木纹、玉石、木地板、烤漆板、壁纸、皮板	Main Materials: White Gravel, Oak, Black Ebony Grain, Jade, Wood Floor, Stoving Varnish Board, Wallpaper, Leather Board

客厅：客厅在色彩搭配上下足了功夫，设计师将黑色的稳重、藏蓝的冷静与银色的高贵融入于整个空间，直线的运用让空间充满情趣，菱格屏风用于装饰电视背景墙，充满创意感。

餐厅：原木色的餐桌保留了明清家具的清雅气度，藏蓝与白瓷餐具营造出宁静的就餐环境。餐厅与厨房相连通，巧妙地以黑色屏风和小吧台分割，趣味横生。

书房：黑茶色为主打色的书房营造出了古朴安静的书香氛围，造型简洁的桌案和博古架弥漫着浓郁的历史气息，简洁而有力度的设计提升了空间的格调。

卧室：卧室的色调雅致，空间极为舒展，通体木纹饰面的墙壁搭配黑色菱格让人心神宁静。飘窗的设计极为精致，两盏古典造型的灯饰仿佛穿越了时空。倚窗而坐品一杯香茗，沉静而优雅。

Living Room: The living room put a lot of emphasis on the color collocations. The designer integrates sedate black color, cold purplish blue into the whole space. The application of lines fill the space with a lot of funs. The diamond lattice screen is applied to the decorative TV background wall, quite innovative.

Dining Hall: The dining table of log wood color maintains the elegant bearing of Ming and Qing Dynasties' furniture. The purplish blue and white china dining accessories produce some serene dining environment.

Study: The study with dark tawny color as the tone creates some primitive and quiet scholarly atmosphere. The desk of concise format and the antique-and-curio shelf send out some intensive historical sphere. The concise and powerful design uplifts the tone of the space.

Bedroom: The bedroom has elegant color tone, and the space is quite stretching in feel. The wall of whole wood pattern veneer is accompanied with black lattice which is soothing. The design of bay window is quite exquisite and the two lighting accessories of classical formats seem to penetrate through time and space. Leaning by the window and tasting a cup of tea, you would find everything tranquil and elegant.

神韵 新中式空间设计典藏

神韵 新中式空间设计典藏

沉浸在陶然时光里
正祥橘郡落地样板A户型

Immersed in the Intoxicating time
Zhengxiang Xiangxie Ballet Show Flat Type A

设计公司：福州林开新室内设计有限公司	Design Company: Fuzhou Lin Kaixin Interior Design Co., Ltd.
设 计 师：林开新	Designer: Lin Kaixin
参与设计：余花	Associate Designer: Yu Hua
项目面积：90 ㎡	Project Area: 90 ㎡
主要材料：大理石、实木、艺术涂料	Main Materials: Marble, Solid Wood, Artistic Coating

本案的设计定位于城市生活与旅游度假之间，既不像东南亚风格那样纯休闲，也不拘泥于传统的都会奢华印象。本案坐落于福州旗山脚下，既远离市区，又与市区有着千丝万缕的联系。它独特的亚市区地理位置、旅游胜地的背景，决定了空间的主题特色和创意焦点。整个空间结构的规划与所处的地域文化脉络相吻合，超越了传统样板房的功能，更容易唤起欣赏者内心的情感呼应。

本案的设计将度假情调与传统文化融汇交融，在传承八闽文脉的同时，又赋予空间简洁明快的现代感，使其保持视觉的连续性和内在气质的统一感，体现深藏于国人内心深处的向往。

The design of this project is positioned between urban life and tourist holiday, not like the pure leisure of Southeast Asian style, nor confined to traditional metropolitan luxurious impressions. This project is located at the foot of Mount Qi in Fuzhou, far away from the urban center, while with some countless ties with it. Its peculiar suburban geological location and background of tourist resort determine the space's themes and innovation focus. The planning of the whole space's structure echoes the local geological cultural context, surpassing the functions of traditional show flat, yet arousing emotional sympathy from the viewers' hearts.

The project design integrates holiday appeals and traditional culture. While inheriting the cultural context of Fujian Province, the design entrusts the space with concise and brisk modern feel, making it maintain the integrity of visual continuity and inner temperament and displaying the longings deep in the people's hearts.

平面布置图

京华烟云
沿海地产赛洛城样板间
A Moment in Beijing
Los Angeles Showroom in the Coastal Real Estate

设计公司：重庆品辰装饰工程设计有限公司	Design Company: Chongqing Pinchen Decorative Engineering and Design Co., Ltd.
设计师：庞一飞、张雁	Designer: Pang Yifei, Zhang Yan
项目地点：重庆	Project Location: Chongqing
项目面积：125 ㎡	Project Area: 125 m²
主要材料：藏青色木作、白色软包、木雕花等	Major Materials: Navy Blue Woodwork, White Soft Roll, Wood Carving

"宅中有园，园中有屋，屋中有树，树上有天，天中有月。"

—林语堂

岁月静好，时光温柔。手握一杯清茶，独自在《京华烟云》里漫步，字里行间，闭上眼睛好似聆听到一曲古乐，寄托着你我的情思。阳光迎窗，古意空间述说着远古的故事。万籁俱寂，清风徐来，在花开的时间里相聚相守。空间如同文字一样，是有灵性的，给予人诗的意境、画的意象。

穿越夜的弥漫，依稀浮现，如梦似幻。窗外又是这个季节，春光又添细雨。镌刻在时光的记忆里，清浅流连，彼此相依，雾积云散，行到水穷处，坐看云起时。

Garden inside the residence, house inside the garden, tree inside the house, sky above the tree, moon in the sky."

–Lin Yutang

Time and tide are so tranquil and gentle. With a cup of tea in hand, wandering lonely in A Moment in Beijing, when you close your eyes, you seem to hear the ancient music among the lines, which carry our endless love. Sunshine shining through the window, the antique space is telling the distant stories. Everything is quiet, the fresh breeze blows gently, and we get together in the time with blossoming flowers. The space is just like characters, with spirituality and entrusting people with poetic artistic conceptions and picturesque image.

Through the pervading night, something is showing up dimly, just like a dream. Outside the window, the season is back, spring has come with some drizzle. Engraved in the memories of time, all is connected with each other, when the fog gathers, the clouds are gone. Walk till the water checks the path, sit and watch the rising clouds.

神韵 新中式空间设计典藏

古韵新律
沿海地产赛洛城样板间
Ancient Charm and New Melody
Yangzhu Garden, Chinese Style Show Flat

设计公司：KSL设计事务所	Design Company: KSL DESIGN(HK)LTD.
设计师：林冠成	Designer: Andy Lam
项目地点：广东深圳	Project Location: Shenzhen of Guangdong Province
项目面积：142㎡	Project Area: 142 m²
主要材料：水曲柳、黑钢、草编壁纸、夹丝玻璃	Main Materials: Ash Tree, Black Steel, Straw Plaited Wallpaper, Wired Glass

端庄清雅、气韵丰华、精致含蓄的东方传统审美情境在本案中得到完美的展现，深沉而优雅的空间气质是本案的亮点。

靛蓝与明黄色相互碰撞，散发出深藏其中的浪漫情调。草编壁纸与木纹地板的搭配诠释出了醉人的古典气质。整个空间无需过多修饰，雕花的瓦当、华美的丝绸、质朴的陶瓷鼓凳、圆润的瓷瓶，看似随意的点缀，却尽显古雅中式的内敛气韵，毫不张扬而又深入心扉。沉静而优雅、低调而充满韵律的室内设计让人心醉。

The solemn, elegant, graceful, artistic, exquisite and restrained traditional oriental aesthetic situation is displayed to the full in this project. The profound and elegant space temperament is the highlight of this project.

Indigo and bright yellow color collide here, displaying the romantic charms deeply hidden inside. The collocation of straw plaited wallpaper and wood grain floorboard interprets the intoxicating classical temperament. The whole space does not need too many decorations, yet the carving eaves tile, grand silk, primitive ceramic stool, round and smooth china vase and other seemingly casual ornaments all display the restrained temperament of classical Chinese style, being low-profile, yet going into people's heart. The solemn and elegant, low-profile and full of charms interior design is just so fascinating.

神韵 新中式空间设计典藏

琵琶曲
天悦湾B栋3号样板房
Pipa Tune
Bay B Building No. 3 room Rosedale

设计公司：史礼瑞设计有限公司
项目面积：83 ㎡
主要材料：木纹石、咖啡木、扪皮、印花镜、手绘地毯、马赛克

Design Company: STEVE.S Design Co. Ltd.
Project Area: 83 ㎡
Main Materials: Wood stone, Coffee wood, Ammonites skin, Printing mirror, Hand-painted carpet, mosaic

本套方案从传统中式装饰中提炼出简约符号，对中国传统文化重新演绎。将中国传统元素与现代人的审美情趣重新结合，融入到现代装饰中，同时注入文化气息，更具韵味和文化内涵，有别于传统的讲究雕琢的中式古典设计，讲究整体效果，在强调主人文化品位和自身修养的同时，更注重生活的舒适性。让鉴赏者感受艺术设计的魅力。

This set of scheme to extract a simple symbol from traditional Chinese styledecoration, a new interpretation of the Chinese traditional culture. Chinatraditional elements with modern people's aesthetic taste to combine. Into the modern decoration, at the same time into the cultural atmosphere, more flavor and the cultural connotation of Chinese classical design, different from the traditional ornate style, pay attention to the overall effect, emphasizing the masters of their own culture and quality of training at the same time, pay more attention to the lives of comfort. Let the appreciation of those who feelthe charm of art design.

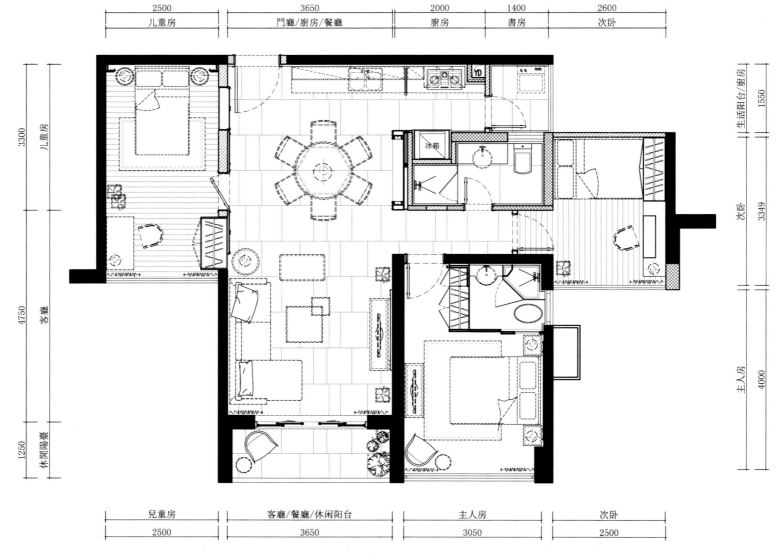

平面布置图

神韵 新中式空间设计典藏

纯美禅意印象
中洲央筑花园样板房
Pure Beauty Zen Impression
Zhongzhou Central Building Garden Show Flat

设计公司：KSL设计事务所	Design Company: KSL DESIGN(HK)LTD.
设计师：林冠成	Designer: Andy Lam
项目地点：广东深圳	Project Location: Shenzhen of Guangdong Province
项目面积：142 ㎡	Project Area: 142 m²
主要材料：水曲柳、黑钢、草编壁纸、夹丝玻璃	Main Materials: Ash Tree, Black Steel, Straw Plaited Wallpaper, Wired Glass

对传统居住空间的热爱，是深藏于中国人内心中的情感，根植在中国人的血液中。本案以黑、白两种清雅的颜色为主调，融入水墨画、中式吊顶、鸟笼、陶瓷、现代中式壁画等装饰性元素，撷取其中的精神内涵，将线条转化为具有装饰意味的形态，使其融会于空间之中，铺陈出全新的中式风格空间。

设计师用现代的设计手法来提炼中式的禅意印象，轻描淡写着宁静高雅的空间感受，在清新素雅的空间氛围里为心灵提供一块净土。

The love for traditional residential space is an emotion deeply hidden in Chinese people's heart and rooted in Chinese people's blood. This project has black and white, two clear and elegant colors as the tone, integrated into decorative elements such as ink and wash painting, Chinese ceiling, bird cage, ceramic tile and modern Chinese mural, etc., whose spiritual connotations are extracted to transform lines into formats of decorative connotations, integrating into the space and producing new Chinese style space.

The designer makes use of modern design approaches to extract Chinese Zen impressions and produces tranquil and elegant space feelings of understatement fashion, providing a pure land for heart inside the fresh and elegant space atmosphere.

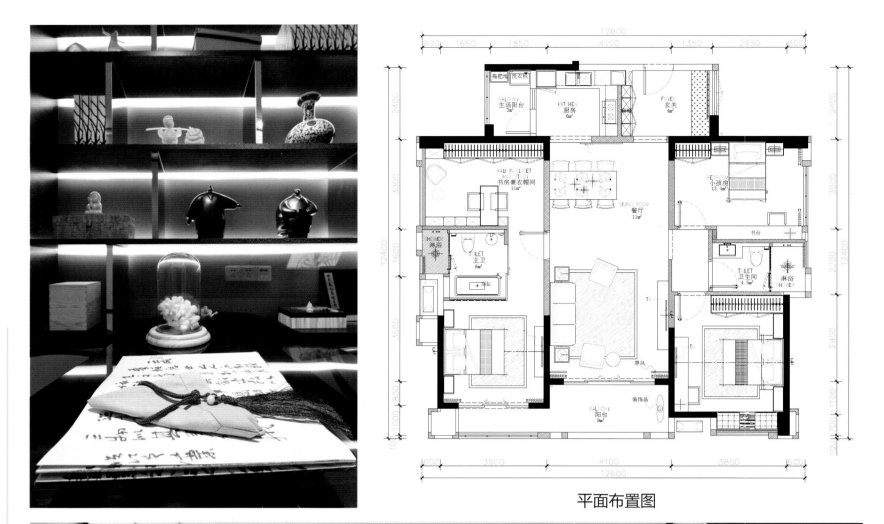

平面布置图

神韵 新中式空间设计典藏

神韵 新中式空间设计典藏

丹青墨影
宁波交通自在城黄宅
Ink Painting Shadow
Ningbo Traffic Comfortable City Huang Zhai

设计公司：金元门设计	Design Company: Golden Gate Design
设 计 师：葛晓彪	Designer: Ge Xiaobiao
项目地点：浙江宁波	Project location: Zhejiang Ningbo
项目面积：140 ㎡	Project Area: 140 m²
主要材料：金丝柚、瓷砖、大理石、壁纸、涂料	Main Materials: Watkins grapefruit, ceramic tile, marble, wall paper, paint

设计师对原有的格局进行了改造，将原来一体化的客餐厅用了"借景"的手法分成两个区域，中间用木色隔断分开，分别在隔断上装饰了两幅花鸟水墨画，不仅利于采光，又很有书香气韵。

餐厅尤其令人印象深刻，以红色为主调，与其他色彩产生了鲜明的视觉反差，传递出一种别样的中式情怀，特别是餐桌旁大红的椅子和靠枕，与草绿色的餐边柜形成对比，彰显出高调的个性。边柜门上刻着的"黄"姓LOGO更是意义非凡，经交流得知，这里90%的家具均由家具设计师自行制图打样并配色。对于这自成一脉的家具，业主黄先生非常喜欢。

客厅毗邻餐厅，设计师用两幅水墨长卷打造的背景墙交相呼应、气势磅礴，有种超凡脱俗的感觉，而黄色与蓝色调的相互映衬无形中衬托了主人的尊贵身份。

卧室分列客厅两侧，儿童房以最具特色的青花瓷为设计主题，以直径2 m的青花瓷盘作为房间的背景，搭配浅蓝色的花鸟壁纸，动静结合、清闲雅致。对面的电视机柜又是一个闪光点，由设计师亲手绘制的丝质壁画和四个瓷盘相互映衬，别有一番韵味。

The designer redesigns the original layout and divides the original unified living room and dining hall through the approaches of "borrowing scenes." The middle area is divided with wood color partition which is decorated with two pieces of paintings of flowers and birds in traditional Chinese style, which facilitate lighting, while displaying abundant scholarly atmosphere.

The dining hall is especially impressive, with red color as the tone, which produces distinct visual contrast with other colors and sends out some different Chinese sensations. The red chairs and bolsters beside the dining table forms contrast with the gramineous side cabinet, manifesting high-profile characteristics. The "黄" logo on the cabinet door is quite significant. After communications, we came to know that drawing, colors, etc. of 90% of the furniture were made by furniture designer. Property owner Mr. Huang is quite satisfied with the furniture of consistent and peculiar style.

The living room neighbors the dining hall and the designer makes use of background wall of two ink painting scrolls to produce magnificent atmosphere, with extraordinary and refined sensations. And the correspondence of yellow and blue colors invisibly sets off the noble status of the master.

The bedrooms are set on both sides of the living room. The children's room has the most peculiar blue and white porcelain as the design theme, with the blue and white porcelain plate of 2m in diameter as the background, accompanied with light blue paintings of flowers and birds in traditional Chinese style, combining active and quiet zones, and being quiet and elegant. The opposite TV cabinet is another highlight, the silk mural painted by the designer himself and four porcelain cups bring out the best in each other, with some different tastes.

神韵 新中式空间设计典藏

神韵 新中式空间设计典藏

幽梦依稀淡如雪
水月周庄某宅
Like Dream Like Snow
One Residential in Moon Water Zhouzhuang

设计公司：萧氏设计	Design Company: Xiao's Design
设计师：萧爱彬	Designer: Xiao Aibin
参与设计：王海婧、蔡娅娟	Associate Designers: Wang Haijing, Cai Yajuan
项目面积：331 ㎡	Project Area: 331 m²
主要材料：原木、柚木、藤编、锈石	Main Materials: Log Wood, Teak Wood, Rattan Plaited Articles, Rust Stone

本案所处的楼盘景色优美，本设计一方面保留了传统东南亚风格的元素，另一方面加入现代材料，将东南亚的禅意与现代的时尚融为一体。

进门就是敞开式的西厨，设计师利用统一的顶部饰面强化西厨与门厅的空间关系，使面积不大的空间借由"分享"的空间来扩展视野。透过纱幔若现的禅意雕像静立在客厅的主入口。步入下沉式的客厅，阳光投射进来，树影婆娑，芭蕉树影透过纱幔投射到地面，安谧的氛围尽现眼底。通过餐厅大面积的落地窗能看到庭院的景色，呼应室内的盆栽，形成自然写意的生活情境。

本案注重建筑内部与外部环境的衔接。在通风采光得到优化的同时，格栅、纱幔的围合遮挡又确保了可放松身心的空间所必备的私密性。在装饰材料上应用原生态的木饰面及文化石、砂岩石，搭配纱幔、棉麻布艺等，尽可能拉大材质间的对比，营造出静穆平和的禅意空间。

The property where this project is located has top landscape, where one can enjoy the mountains and the water. On one hand, the design maintains the elements of traditional Southeast Asian style, while on the other hand, with additional modern materials, Zen style of Southeast Asia and modern fashion are integrated as a whole.

Upon entering the door, you would find the open-style western kitchen, while the designer makes use of integrated ceiling veneer to strengthen the space relationship between western kitchen and parlor, making the seemingly small space extend the visions with "borrowed" space. Through the curtain, you can find the Zen style sculpture quietly standing at the main entrance of living room. Upon stepping into the sunken living room, you can perceive tranquil atmosphere everywhere, sunshine shining inside, shadows of trees dancing in the breeze and shadows of Chinese banana trees cast onto the ground. Through the dining hall's large area French window, you can observe the views of the courtyard, echoing the interior potted plants, performing natural living situations.

This project emphasizes on the connection of interior and exterior environment. While optimizing ventilation and lighting, the enclosing of lattice and curtain can maintain the privacy of a space where people can get relaxed. The decoration materials apply primitive wood veneer, cultural stone and sandstone, accompanied with curtain, linen fabrics, etc., greatly extending the contrasts of materials, while creating solemn and peaceful Zen style space.

云罗雁飞江南秋
水月周庄某宅

Jiangnan Qiu Yun Luo Yanfe
One Residential in Moon Water Zhouzhuang

设计单位：萧氏设计	Design Company: Xiao's Design
主创设计：萧爱彬	Chief Designer : Xiao Aibin
参与设计：詹永林、肖海鹏、王立新	Co-designer: Zhan Yonglin、Xiao Haipeng、Wang Lixin
项目地址：水月周庄	Project Location : Moon Water Zhouzhuang
项目面积：610㎡	Project Area: 610㎡
主要材料：橡木饰面、手绘壁纸、金箔、西班牙	Main Materials: oak veneer, hand-painted wallpaper, gold foil, Spain

地处江南水乡的水月周庄，从布局到建筑均依循地域地貌，不落俗套、粉墙黛瓦。室内设计是建筑环境的延伸，虽定位于新古典气息，但完全摆脱了传统的样式，地面没有繁复的铺装，素色的勾边、单纯的米灰，使之显得高雅且大气。

入口的天花，自由地转换到顶面，墙面及隔断的样式非常独特，有白色的、有木色的，其造型给人深刻的印象。

灯是设计师最喜欢运用的装饰，从本案的灯具设计就能看出来，一个像如意一样的吸顶灯，上面嵌上了荷花。当灯点亮时，就像映在水中的倒影，充满东方的情愫。

室外水中的荷花与室内的山水、花影相印成趣，室内外交融在一起。建筑上的缺点被设计师一个个化解，变成了作品的亮点。

Moon Water Zhouzhuang located in the Yangtze River Delta follows regional landscape, unconventional, slim and graceful, white walls and black tile from the layout to the building. Interior design is an extension of the building environment. Positioned in the neo-classical flavor, but it totally get rid of the traditional style. There is no complicated pavement on the ground. The plain crochet and simple grey beige looks elegant and atmospheric.

Entrance ceilings converts to the top surface freely. The wall and partition style in white and wood color are very unique, which give us a deep impression.

Light is the designer 's favorite decoration which can be seen from the lighting design of this case. A ceiling lamp as Ruyi is inlaid the lotus. When the lamp is lit, it looks like the reflection in the water, fulling of the oriental feeling.

The outdoor lotus in the water and the indoor landscape and shadows contrast finely with each other. The disadvantages of the building are resolved by the designer one by one, and the shortcomings became the highlights of the work.

神韵 新中式空间设计典藏

人境外，听春禽
恒信.春秋府D户型样板房

People Overseas, Listen to the Spring Bird
Hanson• spring and Autumn House D+Showflat

设计单位：深圳市盘石室内设计有限公司 吴文粒设计事务所	Design Company: Shenzhen Panshi Max Interiors Co. Ltd.,/Wu Wenli Design Office
设 计 师：吴文粒、陆伟英	Designer: Wu Wenli、Lu Weiying
参与设计：陈东成	Participate Designer: Chen Dongcheng
项目地点：荆州	Project location: Jingzhou
项目面积：125 ㎡	Project Area: 125 ㎡
主要材料：蓝金砂大理石、新月亮谷大理石、多彩灰大理石、罗马洞石瓷砖、墙布、皮革、南美柚木、美国胡桃木实木地板、菠萝格户外木、灰玻、灰镜、亮面黑钢	Main Materials:Valley sand marble,Blue Moon marble, colorful new grey marble, travertinetile, wall cloth, leather, Rome, South America America teak, walnut wood floors, merbau outdoor wood, gray glass, mirror, bright black ash

东方艺术的最高境界是写意，追求挥洒自如、一气呵成和苍劲质朴的艺术造诣。在本案例中，设计师意图达到同样的境界，传达大气、质朴及优雅的空间氛围。本案将东方传统的气韵蕴藏在室内空间中，通过中国传统建筑中轴对称的平面布局和立面构图，以及传统的中国红、中式屏风、精心设计的原创家具，使空间呈现出现代中式空间的大气质感。

室内处处都摆放着设计师精心设计的陈列品，如家具、灯饰和挂件，陈列着各种精致的中国物件，这些物件与现代的设计在恰当的空间邂逅，让现代设计蕴涵着深深的中国历史文化印迹，使古老的中国艺术焕发出活力。设计师对中国元素准确的把握、运用和组合，塑造出一个具有东方文化底蕴，又兼具现代、时尚的样板空间。在传达尊贵大气的同时，让人切实地感受东方的人文古韵。在弘扬中华深厚文化底蕴的同时，颠覆传统，注入新的元素，使空间透露出中式的禅意和朴实，使空间的每个景致和角度都使人感觉到喜悦，使人在视觉上享受到一种"寥寥人境外，闲坐听春禽"的悠闲境界。

The highest realm of Oriental art style, the pursuit of write and draw freely as one wishes, a synthetic and vigorous and simple artistic attainments. In this case, the designer intended to achieve the same level, communication space atmosphere, simple and elegant atmosphere. The traditional Eastern spirit contained in the interior space, through China traditional architectural symmetry plane layout and facade composition, the entire space using the traditional Chinese red, continuation of traditional Chinese style screen, and original furniture design, showing the modern Chinese space atmosphere.

Indoor everywhere decorated with Designer displays, such as furniture, lighting and accessories, all kinds of exquisite display China objects, these objects and modern design in the space encounter appropriate, make modern design contains deep China historical cultural imprinting, let old Chinese artistic vitality, grasp, use and composition the designer of the Chinese elements accurate. An oriental cultural background, has now, fashion model space. In conveying the distinguished atmosphere at the same time, let a person to feel Oriental cultural relics. Subversion of the traditional culture in developing at the same time, the deep, inject new elements, revealed the Chinese Zen and the simple Wenxin elegant space, each scene and angle space makes people feel the joy. The visual, another kind of enjoyment, to achieve "a few people outside, sit listening to the spring bird" realm.

神韵 新中式空间设计典藏

平面布置图

墨语
深房传麒尚林1栋B4样板房
Ink language
Legend Scenery Building 1 B4 Show Flat

设计单位:福建国广一叶建筑装饰设计工程有限公司	Design Company: Fujian Guoguang Yiye Architectural Decoration Engineering co. Ltd
方案审定：叶斌	Program Validation : Ye Bin
设计师：金舒扬、刘国铭、陈剑英、李宏、王其飞、蔡加泉、张慧晶、余峰	Designers: Jin Shuyang, Liu Guoming, Chen Jianying, Li Hong, Wang Qifei, Cai Jiaquan, Zhang Huijing, Yu Feng
项目地址：福州三坊七巷	Project Location : Seventh Alley Three Lane Fuzhou
项目面积：4000 ㎡	Project Area: 4000 ㎡
主要材料：青石板、仿古砖、乳胶漆、方管、杉木做旧	Main Materials : quartzite, antique tiles, latex paint, square tube, do the old fir

久居都市的人们都希望得到心灵的舒缓和灵魂的寄托，朴实自然的居室风格可以让人暂时忘掉所有的烦恼和忧伤，给人带来亲近自然的快感。本案的平面布局开敞且大气，空间中线与面相互穿插，注重空间的交互性及环境的融合。

设计师以"墨"为空间主题，利用传统的书画、静物水墨画作为空间元素，配以方正有力的线条分割和山水纹路的石材，使整个空间充满中国古典文化所特有的沉稳、大气感。色彩设计在沉稳的黑色调的基础上，融入部分中国传统色彩中的正红色与宝蓝色，为沉稳的空间锦上添花，不至于使人感觉乏味、枯燥。

People who dwell in metropolis always want to get psychological relief and psychic reliance. Primitive and natural residential style can make people forget about all the worries and distresses momentarily, creating for people pleasant sensations close to nature. The plane layout of this project is broad and magnificent, with space center lines and surfaces interlacing alternately with each other and focusing on space interactions and environmental integrations.

The designer has "ink" as the space theme and makes use of traditional painting and calligraphy, and still-life ink and wash painting as the space elements, accompanied with square and forceful line segmentations and stones of integral landscape vein, which makes the whole space be filled with the sedateness and magnificence exclusive to Chinese classical culture. Based on the sedate black color tone, the color design integrates the bright red and Klein blue of traditional Chinese colors, adding brilliance to the calm space, while not making people feel tedious or bald.

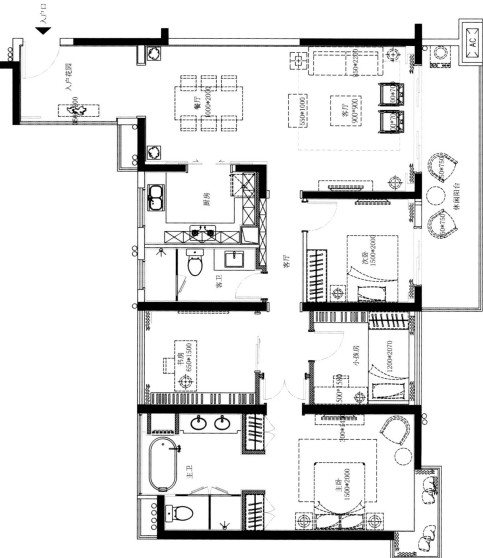

平面布置图

老洋房的海派风情
上海滩花园洋房

Old villa Contains the Shanghai Style
Shanghai Beach Garden Villa

设计单位：全筑设计	Design Company: Trendzōne Design
设 计 师：陆震峰	Designer: Lu Zhenfeng
项目面积：320 ㎡	Project Area: 320 ㎡
主要材料：大理石、壁纸、地毯、木饰面、乳胶漆	Main Materials: Marble, Wallpaper, Carpet, Wood finishes, latex Paint
摄影：三像摄 张静	Photography: Three Photography Zhang Jing

本案展现的是上海20世纪30年代老洋房的海派风情，设计师以古朴大气的格调表现当时老上海的雍容华贵。设计师特别选用最具海派风情的老上海石库门的造型，在客厅、餐厅处均采用高木制护墙板的做法，过渡空间以木制的欧式回廊来表现，墙面正中配以木质壁炉。顶端的装饰进行了多层次、多角度的折边，以水晶吊灯和大理石嵌块来延续上海老洋房黑白相间的装饰主题，并以老式留声机、精美绣品和黑色真皮沙发，以及古韵悠存的瓷器、风扇等来展示精致的海派生活，展现厚积而薄发的海派文化底蕴。

The case shows the old house amorous feelings style in Shanghai in the 1930s, see the elegant of old Shanghai through the quaint and majestically style. Designer selects the Shanghai Stone Gate styling with most Shanghai amorous feelings, uses high wooden wainscoting in the living and dining room, show transitional space by the European corridors of the wooden, and matches wooden fireplace in the middle of wall. At the top of the decoration carries on the multi-level multi-angle folding, goes on the black and white decoration theme of old house in Shanghai by using crystal chandeliers and marble inlay. And old-fashioned gramophone, exquisite embroidery, black leather sofa, ancient porcelain and fan show the exquisite life and profound Shanghai-style culture.

神韵 新中式空间设计典藏

地下一层平面图

一层平面图

二层平面图

三层平面图

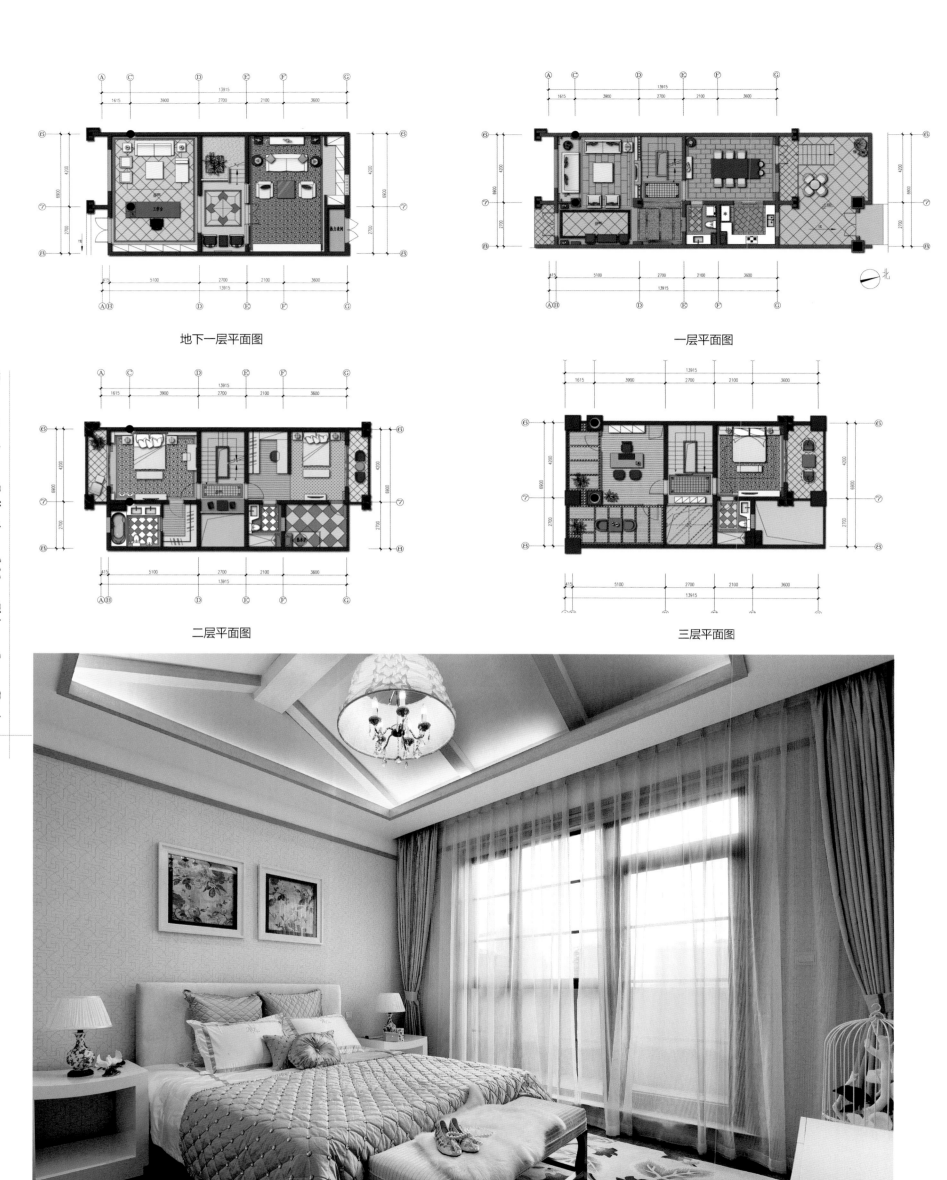

神韵 新中式空间设计典藏

山水有禅意
盛世嘉园某宅
Landscape Contains A Zen
A Residence in Shengshi Jiayuan

设计公司：辉度空间设计	Design Company: Huidu Space Design
设计师：夏伟	Designer: Xia Wei
项目面积：130 ㎡	Project Area: 130 m²
主要材料：雕花板、不锈钢、硬包、壁纸、烤漆玻璃、马赛克	Main Materials: Carving Plate, Stainless Steel, Hard Roll, Wallpaper, Stoving Varnish Glass, Mosaic Tile

佛教中的禅讲究的是虚灵宁静，摒弃外在环境，不受其影响。中国山水画的缘物寄情，托物言志，成为中国花鸟画的灵魂和本质。

春秋的时候，读书人便喜欢在厅堂的案几上东向置瓶，西向摆镜，取义东瓶西镜，名为"平静"。后世，更有白居易"人间四月芳菲尽，山寺桃花始盛开"的"心境"以及陶渊明笔下"桃花夹岸，落英缤纷"的桃源美景写照。其上种种所追求的无非是远离世事的正、清、和、雅。

在人心浮躁的今天，追求心灵的平静与宁和仿佛是人们的一种追求，是只有在繁华过后才能拥有的从容。既然出门在外的拼搏必不可少，那至少在回到家后，让我们放下一切，放松心灵，与家人共享一份静谧和清闲。

Zen in Buddhism stresses tranquility, abandons the outside things and shall not get influenced by them. For Chinese landscape paintings, they apply objects to display the emotions and ambitions, which is the soul and essence of paintings of flowers and birds in traditional Chinese style.

In Spring and Autumn Period, the intellectuals would like to set the vase on the east side of the desk inside the living room and mirror on the west, with connotations of east vase and west mirror, with the assonance of "tranquility". Thereafter, there is the mental states of poet Bai Juyi "In April of the mundane world, all the flowers are gone, yet the peach flowers of mountain temple are just starting to blossom," and the peach garden portrayal by Tao Tuanming, "The peach flowers blossoming on both sides of the river, fallen petals scattering and flying around like snow flakes." What the above pursues are the positive, clear, harmonious and elegant atmosphere far away from the mundane world.

In today's world, people's hearts are turbulent. It seems to be one pursuit of people to seek for psychic tranquility and serenity. It is the easiness that people can own only after prosperity. Now that the struggle outside home is necessary, when we go back home, let's put everything down, get relaxed and enjoy the tranquility and leisure with the family members.

神韵 新中式空间设计典藏

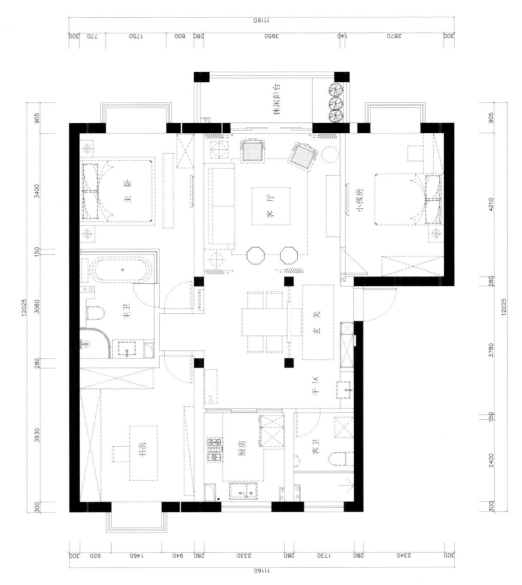

平面布置图

陶瓷智慧与刺绣文化
鹤山十里方圆

Ceramic Wisdom and Embroidery Culture
He Shan "Ten Miles" Villa

设计公司：尚策室内设计顾问（深圳）有限公司
设计师：李奇恩
项目地点：广东鹤山
项目面积：200 ㎡
主要材料：拼花地板、磨砂玻璃、木墙

Design Company: Shang Ce Interior Design Co., Ltd.
Designer: Li Qi'en
Project Location: He Shan, Guang Dong
Project Area: 200 ㎡
Main Materials: Block floor, ground glass, wooden wall

本案想探索一种适合现代中国人居住，又能容纳西方文化的设计。首先，提取代表中国文化的陶瓷元素，以西式的装饰方法挂在墙壁上，白色的餐桌和餐椅，配置秀气大方的软装与插花，保留了中国人灵动清雅的姿态与气韵。磨砂玻璃的敞门，既是隔断，又是艺术的体现。玻璃敞门上的刺绣，让人可以欣赏到中国的刺绣文化。书房里用花团锦簇、构思细腻、颜色华丽的壁纸装饰，表现出中国文人诗画兼通的情怀。

We try to design a house that contains elements of western culture for contemporary Chinese people. The decoration on the wall combines Chinese ceramic elements and western style. On the white dining table and chairs are the delicate soft-mounted elements and flower arrangements, which represent an elegant and smart air of Chinese culture. The door made of ground glass functions as a partition as well as a work of art. The embroidery element on the door is a representative of China's embroidery culture. Besides, the study room is exquisitely decorated by colored flower-pattern wallpaper that shows a sentiment of Chinese literati.

神韵 新中式空间设计典藏

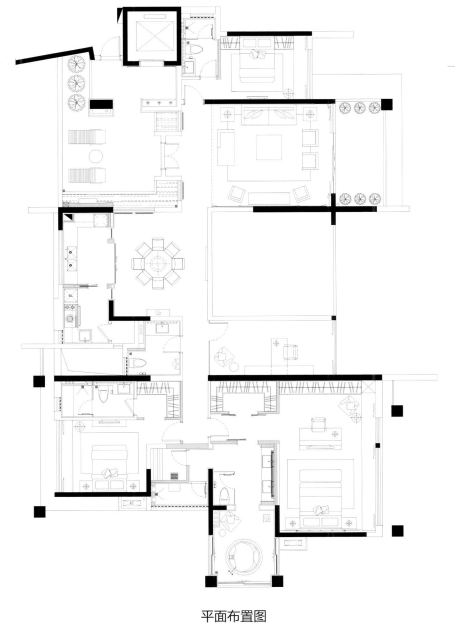

平面布置图

神韵 新中式空间设计典藏

东方神韵 绝色风华
涛景湾豪宅
Oriental Verve, Beauty and Elegance
Taojingwan Mansion

设计公司：深圳锦尚森迪软饰顾问机构	Design Company: Shenzhen Jinshang Sundy Soft Decoration Consultants Institution
设计师：李志艳	Designer: Li Zhiyan
项目地点：广东广州	Project Location: Guangzhou of Guangdong Province
项目面积：200 ㎡	Project Area: 200 m²

新东方风格与现代时尚的完美结合，创造出一个别开生面的居住环境。不同于纯中式的沉静、悠远，区别于现代奢华的高端、华贵，它拥有优雅大气的外观，内蕴沉静、内敛，带给人心底的触动，引起人们最强烈的共鸣。

设计师以干净、优雅的色系为空间氛围做铺垫。简洁而不简单的设计和隐含禅意的饰品，抒发出浓厚的东方意境，描绘出令人惊叹的东方神韵，让人回味。

The perfect combination of New-Oriental style and modern fashion produces a new residential environment. It is different from the serenity and profoundness of pure Chinese style, differs from the high-end quality and magnificence of modern luxury. It has elegant and sublime outlook, with sedate and constrained connotations, bringing to people inmost touch and arousing the most intensive resonance.

The designer applies clear and elegant color system to foreshadow the space atmosphere. The design being concise but not simple and the ornaments of Zen charms send out intensive oriental artistic conceptions and create outstanding oriental charms, arousing people's aftertastes.

神韵 新中式空间设计典藏

禅意东南亚
嘉宝梦之湾
The Zen of Southeast Asia
Jiabao Dream Bay Show Flat

设计单位：上海乐尚装饰设计工程有限公司	Design Company: Shanghai LESTYLE Decorative Design and Engineering Co., Ltd.
项目面积：270㎡	Project Area: 270㎡
主要材料：橡木染色、麻编、质感壁纸、米色理石、木质条纹	Main Materials: Dyed Oak, Linen, Texture Wallpaper, Beige Marble, Wood Stripe

传统的泰式风格喜欢运用浓烈的色彩，地处亚热带的泰国盛产水果，运用水果色系是泰式风格的特点。在这种色彩"热舞"中，舒张中有含蓄、妩媚中有神秘、平和中有激情，能够体现出人们的热情和生活的快乐。

泰式风格的饰品多以器皿为主，造型古朴、雅致，图案以及装饰纹样具有浓郁的民族风情，镂空的门窗、柳藤编制的椅子，都是典型的泰式风格。

本案在整体上吸取泰式风格的轮廓，并将现代人的审美情趣与传统相结合。在整体的硬装及软装设计上，围绕休闲、舒适、禅意三个关键词。泰国也是一个崇尚佛教的国家，在室内设计中注重禅境氛围的营造。设计师想在钢筋水泥构筑的城市之外，营造出一个世外桃源般的热带风情之家。

在装修材料的选择上，主要以橡木染色、原木色木质条纹以及泰式麻编与质感壁纸为主，地面铺设米色理石。

整体配色以木色系为主，蓝色系中的蓝灰、蓝绿及暗红色的色彩点缀，体现出空间的休闲情趣。米色布艺及大体量的家具，突出家具的造型感，同样营造出了休闲旅游般的度假情调。

Traditional Thai style enjoys applying intensive colors. Located in the subtropical zone, Thailand abounds in fruit and fruit color system is a major feature of Thai style. Inside this color hot dance, there are connotations in enthusiasm, mysteries in grace and passion in peaceful moods, which can manifest people's zeal and life joys.

Thai style accessories are mainly utensils, with archaic and elegant format and intensive national charms for the decorative patterns. The hollow-out doors and windows and cane-made chairs are all typical Thai style.

As a whole, this project introduces layout of Thai style, combining modern people's aesthetic interests and traditions. The whole hard decoration and soft decoration design centers on three key words: leisure, comfort and Zen. Thailand is also a country advocating Buddhism, which focus on the creation of Zen style atmosphere in the interior design. The designer aims to create a paradise like home with amorous feelings away from the city of armored concrete.

The selection of decorative materials mainly focus on dyed oak, log wood color wooden stripe and texture wallpaper, etc.

The ground is paved with beige marble.
As a whole, the color focuses on wood color. Color ornaments of blue gray, blue green and dark red in the blue color system represents the leisure interests of the space. Beige cloth and large furniture highlights the format of the furniture, while creating leisure travel-like holiday amorous feelings.

神韵 新中式空间设计典藏

一镜一禅心
泰禾红树林联排别墅
One Mirror, One Zen Heart
Taihe Mangrove Townhouse

设计公司：简艺东方设计机构	Design Company: Jane Arts Oriental design agency
设计师：孙长健、林元娜	Designers: Sun Changjian, Lin Yuan'na
项目地点：泰禾红树林联排别墅	Project Location: CAC mangrove townhouse
项目面积：325 ㎡	Project Area: 325 ㎡
主要材料：楼兰陶瓷、米洛西石砖、柔然壁纸、美兆家具、地板	Main Materials: Loulan ceramic, Marmocer brick, ROEN wallpaper, Maxxa furniture, Macaws floor

"愚人求境不求心，智者求心不求境"，用这句禅语来解释禅心的心境。本案通过平静的水及玻璃反射形成的镜面，在不同的镜之中，看到空间的无限延伸、移步换景、曲径通幽。就如同每个人的心境一样，思维方式的不同决定看事物的角度与立场。在这个作品中，每个镜面都能反射不一样的看点，只有权衡真实和虚幻，才能做到"触目遇缘无障碍"，使心境更加开阔，发掘暗藏空间中的更多亮点！

"The fool for exit for the heart, the heart does not seek the accurate boundary for", Zen Buddhism Heart phrase Chan Yu to explain, Formed in this case through the calm water and glass reflection mirror, Among different mirror, See the continuous extension of the space、a different view with every step/take a step and the scenery will chang、a winding path leading to a secluded place.Just as each person's state of mind, The different ways of thinking determines things different angle positions.Each mirror can reflect different Aspect works, Only be weighed against the real and the unreal, To do "See case of the margin of the accessibility of the accurate" More open mind, Explore more highlights hidden space.

神韵 新中式空间设计典藏

写意东方 古韵今生
大儒世家卧虎2#305
To Freehand Oriental Rhyme Life
Daru Family Crouching Tiger 2#305

设计单位：广州华浔品位装饰	Design Company: Guangzhou Hua Xun Taste Decoration Company
设计师：黄育波	Designer: Huang Yubo
项目地点：福州	Project Location: Fuzhou
项目面积：120 ㎡	Project Area: 120 ㎡

功能：为了很好地展现公共空间，设计师在很多地方都应用了玻璃材质。客厅、餐厅、入户花园是一个整体的空间；书房和厨房形成一条轴线；主卧和书房是一个盒子式的设计，在书房和主卧之间设计了一道隐形门，关上门可形成两个独立空间，打开门可成为同一个空间。

材质：在材质应用上，设计师以绿色、环保、节能为前提，一方面使得空间的结构感增强，另一方面使空间的环保系数提高。电视背景墙和八角门套采用免漆雅典柚木木皮饰面，房间的地板采用软木地板，可以实现资源的循环利用。

灯光：在灯光设计方面，设计师采用2700K色温的光源，使墙壁的蒙托漆和雅典柚木木皮的质感更好地表现出来。

风格：空间的装修风格和装饰细节既有现代简约的利落，也有中式传统的韵味。于是在这个新东方意境的家居环境中，创意与功能兼得，现代与中式并存。

色彩丰富的油画与质朴的仿古砖形成鲜明的视觉反差，采用实木木皮构建的中式八角窗与现代风格的桌椅形成独特的搭配，颇有朴素中见时尚的意味。

Function: In order to show the public space better, we use glass materials in many places. The living room, dining room, home garden is a whole space. Study and kitchen form an axis, and the door is made by glass. The master bedroom and study is a box-type design. An invisible door set between the study and the master bedroom. When the door closed, there are two independent spaces; open the door it is a whole space.

Material: For materials, we premise on environmental protection and energy saving. We have two considerations in the choice of materials. On one hand, to enhance the space structure, on the other hand, to improve environmental protection factor. TV backdrop and star anise door cover use non-painting Athens teak wood veneer; the room floor is made of cork, it is the renewable resource.

Lighting: For the design of lighting, the 2700 k color temperature light was used to come out the texture of the Meng Tuo lacquer and Athens teak veneer on the wall.

Style: The decoration style and details of the space are with both modern minimalist convenience and Chinese traditional flavor. Therefore, creativity and function, modern and Chinese style both coexist in this new oriental home environment.

Colorful paintings and rustic antique brick form the stark visual contrast. The collocation of the solid wood Chinese octagonal windows and modern tables and chairs seems to see fashion from simple.

平面布置图

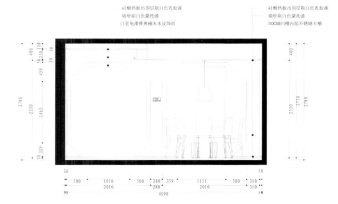

厨房入户门八角门立面图

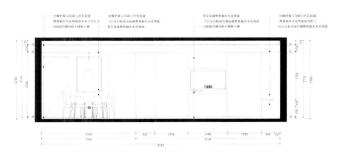

电视背景墙立面图

The Third Chapter
Sales Offices Space

第三章
售楼处空间

密不透风、疏可走马
苍海一墅售楼部
Dense Yet Sparse
The Villa in Aomi Sales Department

设计单位：重庆品辰装饰设计有限公司	Design Company: Pin Chen Decoration Co., Ltd.
主设计师：庞一飞	Master Designer: Pang Yifei
项目面积：2000㎡	Project Area: 2000㎡
主要材料：古木纹大理石、锈石、黄花岗石斧垛面、水曲柳、索色木地板	Main Materials: wood ancient marble, stack stone axes Huanghuagang face, Ash, cable color wood flooring
摄影师：张起麟	Photographer: Zhang Qilin

到了云南大理，谁也不能拒绝"下关风，上关花，苍山雪，洱海月"的诱惑。

云南实力房地产开发有限公司开发的旅游地产项目"苍海一墅"，让人联想到现代诗人肖草的《大丽梦》，诗曰："苍山雪，洱海月。龙潜洱海，飞起万片银雪。海浴玉龙，怀抱一轮皓月。洱海映苍山，永烙心中。白水河，彩云月。极目远舒，玉龙挥泪长河。笑看中秋，洱海佩戴新月。大理对丽江，追忆梦索"。

设计师善于把握个案背后的创作意象，在苍海一墅售楼部的设计中，将鱼的元素提炼出来，贯穿到整个设计中。洽谈区独特的灯具上面手绘着大理白族喜欢的纹样，使当地的特色文化在不经意间融入到空间的每一个角落。

大理常有和煦的阳光洒落，设计师利用了光与影的无穷魅力。如同柯布西耶所说："从当地的自然环境中吸取设计的灵感……使光进入建筑的内部环境，将它朴实无华的特点变成诗一般的意境。"

To Dali Yunnan, nobody can refuse the temptation of "Xiaguan's wind, Shangguan's flowers, Cangshan's snow, Erhai's month".

The development of the strength of Yunnan Real Estate Development Company Limited tourism real estate project "Cang sea Shu", reminiscent of the modern poet Xiao grass "Dahlia dream", said: "Cangshan snow, Erhai month. Erhai flying dragon dive, 000 silver snow. Sea bath ERON, embrace a haoyue. Erhai Ying Cangshan, branded the hearts forever. Baishui River, cloud month. As far as the eye can see far Shu, Changhe ERON crying. Laugh at the Mid Autumn Festival, Erhai wearing the new moon. Dali to Lijiang, recalling dream cable".

The designer is good at grasping the creation image behind the case. In the design of the Villa in Aomi Sales Department, the elements of fish are extracted, and throughout the entire design. Above the discuss district 's unique lamps, there are hand-painted patterns the Dali Bai minority favorite, so that the local cultural characteristics inadvertently blend into every corner of the space.

There are always genial sunshine spilled in Dali. The designer uses light and shadow charm. Like Le Corbusier once said, " draw decorative design inspiration from the local natural environment...... the light enter the building interior environment, its plain feature turn into the poetic mood."

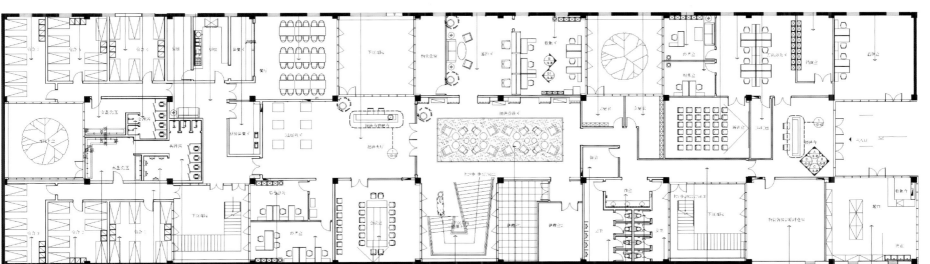

平面布置图

竹意清幽
江湖禅语销售中心
Bamboo Quiet
Rivers and lakes of Zen Sales Center

设计单位：台湾大易国际设计事业有限公司·邱春瑞设计师事务所	Designer Company: Taiwan Dayi International Design Co., Ltd.• Qiu Chunrui Design Firm
设计师：邱春瑞	Designer: Qiu Chunrui
项目地点：江西省宜春市秀江西路206号	Project Location: No. 206 Xiujiang Road Yichun City Jiangxi Province
项目面积：800 ㎡	Project Area: 800 ㎡
主要材料：木纹石大理石、灰麻石大理石、山西黑大理石、凯悦米黄大理石、榆木、水曲柳、肌理木、墙纸、布艺、皮革、乳胶漆、金属、地毯	Main Materials: wood stone, gray granite, Shanxi black, Hyatt beige, elm, ash, wood texture, wallpaper, fabric, leather, latex, metal, carpet

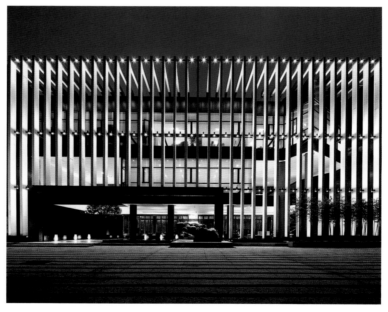

这个设计作品没有过多的装饰，简洁、清秀，处处散发着传统的文化神韵，这就是本案设计的最大特色。项目地处宜春市的"风水宝地"，西边靠近秀江御景花园住宅区，东边毗邻御景国际会馆，南边是湿地公园。销售中心的本体是一家营业多年的海鲜酒楼，进入后厅的就餐区依然能感受到酒楼的气息，这更像是旧楼的再次改造，可以充分发挥设计师的创造力和空间合理再利用的能力。

"室内是建筑的延伸"，建筑和室内不应该是相互独立存在着，而是要相辅相成，这样的认识也是本案的设计基础。整栋建筑分为三层，一层主要是展示空间，其余两层分布着VIP室、办公区和就餐区。通过"里应外合"的串联，使得设计更加富有魅力和吸引力。设计初期，设计师对中国传统合院式的"目"字形的三进院落进行推敲，通过正面的大门后，须穿过一段设计好的水景区域，然后再步入销售中心的正门，这样的设置，能更好的贴合中式传统庭院的婉约感。

格栅是设计的主要元素，纤直的实木线条排列在室内空间中随处可见，寓意着正直、包容、豁达、沉稳。建筑运用钢结构配合栽种的竹子、常青树和人造的水景来营造禅意，浓厚的意境呼之欲出。在设计过程中，设计师始终坚信，传统文化的表达和传递，不能仅仅只拘泥于那些形式上的代表性符号，更重要的是神的塑造和意会。

This design works, not too much decoration, concise, delicate and pretty, everywhere exudes traditional cultural charm, the biggest feature is the case design. The project is located in Yichun City, "Feng Shui", near the West River scenic area show of residential garden, adjacent to the South East Garden International Club, is a wetland park. Sales center of the body is an operating years of restaurants, dining area after entering the hall can still feel the breath of the restaurant, it is more like the old building re transformation, can give full play to the ability of designer creativity and reasonable space reuse.

"The interior is extension of the building", architectural and interior should not exist independently of each other, but to complement each other, this understanding is the design basis of the case. The whole building is divided into three layers, one layer is the main exhibition space. Through the "collaborate from within with forces from outside" series, which makes the design more charming and attractive. The early stages of design, designers Chinese three courtyards of traditional courtyard style "eye" shape of the study, through the front door, Front Gate then entered the sales center, such settings, can better fit the traditional Chinese courtyard subtle feeling.

Grid is the main element of design, fiber straight solid lines are arranged in the indoor space can be seen everywhere, moral integrity, The use of steel structure building with planted bamboo, evergreen trees and artificial lake to create a Zen, strong artistic conception of be vividly portrayed. In the design process, designers always believed, and transfer the expression of traditional culture, the representative not only adhere to the forms of symbols, the more important is the shape of the deity and understanding.

神韵 新中式空间设计典藏

神韵 新中式空间设计典藏

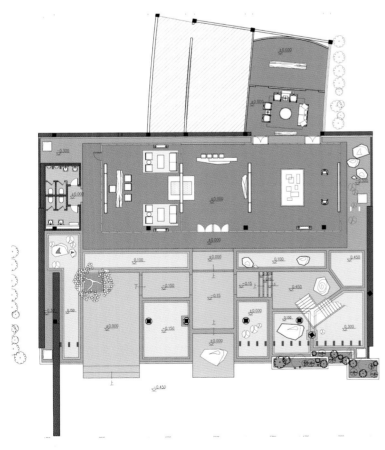

一层平面图

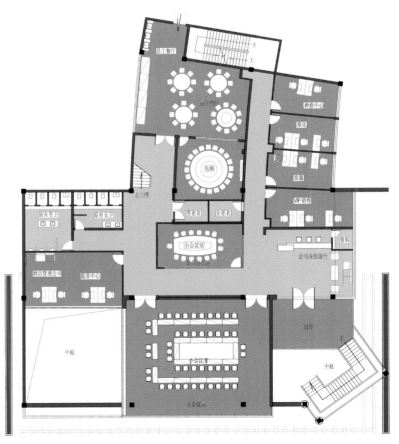

二层平面图

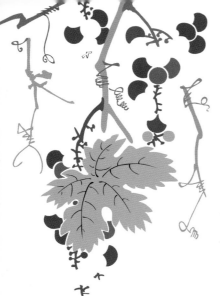

悠悠古香 淡淡神韵
五洲世纪城售楼处设计
Long Ancient Incense, Subtle Charm
Wuzhou Century City Sales Offices Design

设计单位：朗昇空间设计	Design Company: Lonson Space Design
项目地址：清远	Project Location: Qingyuan
主要材料：大理石、地板胶、涂料、玻璃、防火板等	Main Materials: Marble, Floor Glue, Paint, Glass, Fireproof Board etc.

售楼处建筑外观设计采用岭南建筑风格，简洁典雅，独具特色。园林设计亦具岭南园林特色，小巧精制，绿意盎然。青砖、灰瓦、雕刻、花格、古树、喷泉，这一组充满着悠悠古香的中式元素，描绘了一幅具有淡淡神韵气质的中国山水画。为保持与建筑外观设计、园林设计风格协调一致，设计师在室内空间的设计上，亦采用新中式设计的手法，运用现代材料与中式家具、屏风、瓷器、木雕等古典元素，将空间的大气及东方神韵，完美地表现出来，使客人在购房时轻松地感受新中式的人文气息。

大堂设计利用建筑原有的高挑结构，经过精心设计后，显得气势恢宏，光彩熠熠。金属铝格栅顶棚从顶部纵横组合，倾泻而下，使上、下空间具有层次感和延续性。从天花中心悬吊下来的一组巨型中国结水晶吊灯，质感华美富丽，照射着巨大的沙盘展示区，与一侧的组团沙盘区隔邻相应。宽大的售楼处接待台大方、实用，其背景墙利用石材自然的纹理形成一幅中式山水画作，充满着巧夺天工的设计效果。两侧的几组古典灯具和陶罐饰品，更加锦上添花。透过透明的外墙玻璃，可见外面美丽的园林，而与其一墙之隔的大堂内侧，一边有序摆放的几组圆润光洁的陶罐，与外景一起成为亮丽的风景线，渲染了大厅内的中式气息。而另一边则摆放了几组博古架，在摆放精美饰品的同时，还摆放了供楼盘宣传使用的材料，以兼顾到空间的实用性。

竖立在大堂正面的是几幅巨型的中式雕花屏风，由精雕细刻的金属制作，富中式神韵，又具新意。通过中式屏风，既将一、二楼空间有机连接在一起，同时将大堂与洽谈区两个空间恰当地间隔开来。洽谈区分为户型展示区、洽谈区、签约区和水吧服务区等，整个空间相互交汇、通透、开阔、层次感丰富。美丽的落地窗旁，有多组浅蓝、浅灰色的组合茶几沙发，在天蓝色的窗帘及古典壁灯的映衬之下，共同营造出温软舒适、色彩丰富的洽谈氛围，以适合销售人员与客户沟通的需要。

洽谈区中心位置则摆放了一长一方两组中式桌椅，围绕长桌摆放了两组太师椅，其夸张的椅背更加引人注目。从洽谈桌上方天花下垂的数盏鸟笼吊灯投射出柔和的灯光，给人以舒适温和的感受。洽谈区两侧的位置还设计有涓涓流水、青青小荷。在流水小荷上方的墙面上，装饰用的锦鲤小品，因其有序同向的造型摆放，似有慢慢游弋的动感，甚具意味。不经意设计的中式庭园意境，似乎更增添了洽谈区悠悠古香、淡淡神韵的中式气息。

位于售楼处两侧还设计有儿童游乐室、音影室、销售办公区、洗手间等。充满着童真趣味的儿童游乐室，天花上的蓝天白云、墙面上的立体森林、沙发上可爱的小动物、城堡滑梯等，其色彩丰富、清新活泼，使在此嬉戏的孩子们可以获得一次难得的、有趣的游乐体验。

值得一提的是洗手间设计，也使用了新中式风格的设计手法，细节设计上一丝不苟。过厅背景墙上又一面巨大的茶色衣冠镜，镜前摆放一组中式圈椅和案几，周围使用大幅梅花点点的壁纸装饰，空间虽微不足道，但中式风韵融贯其间，为如厕的人们提供了一处清雅宁静、心情愉悦的过渡空间。

Sales offices building designs using the south of the Five Ridges architectural style, simple and elegant, unique. Landscape design also with south of the Five Ridges garden features, compact and refined, the greenery. Green, gray tiles, carved, lattice, trees, fountains, this group is full of long one of the Chinese elements, draws a picture with the subtle charm temperament Chinese landscape painting. Design, architecture and landscape design style and appearance to maintain consistent, in the interior space design designer, also adopts the design of the new Chinese style tactics, the use of modern materials and furniture, screens, porcelain, wood carvings and other classical elements, the space atmosphere and Oriental verve, perfectly, make the guest feel relaxed new Chinese humanistic spirit in the purchase time.

The lobby design using the original structure of the building is tall, carefully designed, is magnificent, shine. Metal aluminum grille smallpox from the top aspect composition, coming down, so that the upper and the lower space with a sense of hierarchy and continuity. Fall down a group of giant crystal chandeliers hung from the ceiling China node center, gorgeous gorgeous texture, radiation huge sand table display area, at the same time, with the side of the group sandplay area adjacent the corresponding. Sales offices in big reception and generous, practical, the background wall with stone natural texture form a Chinese landscape painting, full of Art beats nature. design effect. On both sides of the several groups of classical lamps and pottery ornaments, more icing on the cake. Through the transparent glass wall, visible outside the beautiful gardens, with the wall of the inner side of the lobby, orderly placement of several groups of rounded and smooth ones, and location to become beautiful scenery line, rendering the hall Chinese style atmosphere. And the other side placed several groups of shelf, placed in the exquisite jewelry at the same time, also placed for sale promotional materials used, taking into account practicality space.

Erected in the lobby front several giant Chinese style screen, made of carved metal, rich Chinese charm, and innovative. The Chinese style screen, is to make the one or two floor space organically connected together, and the two space lobby and discussion area appropriately spaced. Negotiate divided into apartment layout display area, negotiations, signing area and the water service area, the space between the intersection, transparent, open, rich sense of the

level. Beautiful French window, a plurality of groups of light blue, light gray combination coffee table, under a sky blue curtains and classical wall against the background, and jointly create the soft warm, rich colors of the negotiations atmosphere, to fit the need of the sales staff to communicate with customers.

Discussion area center position is placed a long side of the two group of Chinese style tables and chairs around the table, placed two group chair, the exaggerated back more attract sb.'s attention. From the negotiating table above ceiling drooping hanging down from the beacon cage lamp cast a soft light, give a person with comfortable and gentle feeling. The positions of both sides negotiate district is also designed with trickling water, green lotus. In the water lotus of the wall above, ornamental Koi pieces placed orderly, because of the same to the shape, like a slowly swimming movement, the very means. Casual design Chinese garden artistic conception, seems to add more discussion area leisurely ancient incense, subtle charm of traditional Chinese art.

In the sales offices on both sides is also designed with children's recreation room, sound studio, sales office area, Restroom etc.. Full of innocence interesting children's recreation room, ceiling, wall of blue sky and white clouds on the stereo forest, the sofa cute little animal, Castle slide, its color is rich, fresh and lively, the temporary in children given a rare, interesting recreation experience.

It is worth mentioning that Restroom design, design techniques to use the new Chinese style, be strict in one's demands on detail design. The hall background wall of a huge brown clothes mirror, mirror placed before a group of Chinese chair and desk, a little used around the plum blossom wallpaper decoration, space is not worth mentioning, but Chinese charm coherence in the meantime, gives out a very elegant quiet over space, happy people for toilet.

神韵 新中式空间设计典藏

东南亚的新东方主义
清澜半岛销售会所
The New Orientalism in Southeast Asia
Qinglan Peninsula Club

设计单位：萧氏设计	Design Company: Xiao's Design
设计主创：萧爱彬	Creative Design: Xiao Aibin
设计团队：张想、何侦辉	Design Team: Zhang to He Zhenhui
软装陈设：姚莉慧、郭丽丽	Furnishing: Yao Lihui, Guo Lili
项目地址：海南文昌	Project Location: Hainan Wenchang
项目面积：2000 ㎡	Project Area: 2000 ㎡
主要材料：木饰面、芬兰木、棉麻硬包、大理石、透光云石、藤编	Main Materials: Wood veneer, Wood, Cotton in Finland hard package, Marble, Transparent marble, Rattan

所谓半岛即伸入海面的陆地，清澜半岛270°环视海面，拥有美丽的海景。本建筑为东南亚风格，室内设计以此风格为基础进行延伸，运用了新东方主义的设计理念。

会所初期作售楼之用，完成使命后可能要成为纯粹的接待空间。现有的沙盘将来可以撤掉，中厅可以作为大厅使用。室内设计与建筑设计不同的一点是设计师往往从使用者的心理感受入手，即从心理学方面考虑，空间才有可能做得更好。设计师在入口处做了一个压缩的长廊，这是设计师处理空间的一贯作风，先抑后扬，不让客人一到入口就一目了然，而是曲折迂回，然后豁然开朗。

销控台，不能放在入口，入口处可以安排一个笑容可掬的迎宾小姐，这样符合消费者的心理，客人才愿意进去。销控台虽放在第三空间，但当客人经过入口长廊进入沙盘区后，销控人员即可迎接上去，这样的服务方式才是最好的，客人才不会感到唐突。

朝南的一线海景，与入口的长廊形成对称布置，成为"洽谈区"。这是当客人欣赏完房间、看好楼盘信息后的落脚地，也是开发楼盘的目的所在。将最好的位置、最优美的海景、最舒适的沙发和座椅安排给客人，一边欣赏美景，一边品茗，合作自然容易成功。家具和配饰都是设计师为此空间量身订做的，椅子是主设计师刚获奖的作品，用在清澜半岛中与空间格调相得益彰。萧氏设计把行为心理和销售心理都完整地体现在空间中，这样销售的成功率一定会达到预期的效果，事实也是这样。

The so-called peninsula that extends into the sea land, Qinglan Peninsula 270 degrees around the sea, has the beautiful sea view, the building is a Southeast Asian style, this style of interior design for the extension, the use of the design idea of the new orientalism.

The club to do the initial sales purposes, after completion of the mission may have to be pure reception room. The existing sand can be removed in the future, can use as hall hall. from the perspective of Feng Shui, designers must consider from the psychology, the space it is likely to do better. Entrance had a compression of the gallery, this is the designer of processing spatial consistent style, xianyihouyang, do not let a guest to the entrance will stick out a mile, but full of twists and turns, and then click into place.

Sell control table, can not be placed in the entrance, the entrance can arrange a beam fairly hostess, that accords with consumer psychology, the guests want to go. Pin console though placed in the third space, but when the guests through the corridor into the area, pin control personnel can meet up, this mode of service, is the best, the guests will not feel abrupt.

South of the harbour line, form a symmetrical arrangement with the entrance to the gallery, become "talks". This is when the guest room, settled after enjoying good market information after development projects, is also the purpose of. The best position, the most beautiful sea view, the most comfortable sofa and chairs for the guests, while enjoying the scenery, Shaw's design to the psychology and the sales psychology completely embodied in the design space, this success rate will achieve the desired results, the fact is that.

神韵 新中式空间设计典藏

神韵 新中式空间设计典藏

神韵 新中式空间设计典藏

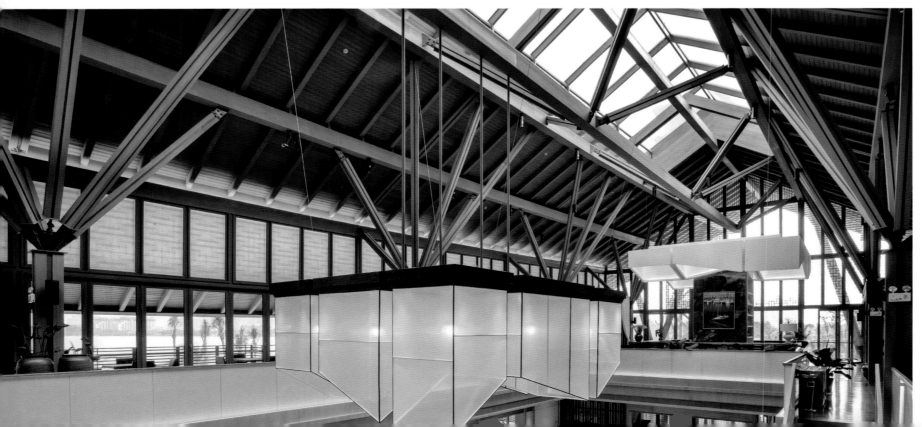

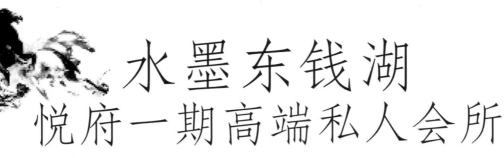

水墨东钱湖
悦府一期高端私人会所
Ink Painting of Dongqian Lake
A High-end Private Club in Yue Fu

设计单位：深圳市昊泽空间设计有限公司	Design Company: Haoze Space Design Co., Ltd.
设计师：韩松	Designer: Han Song
项目地点：浙江省宁波市	Project Location: Ningbo in Zhejiang Province
项目面积：850 m²	Project Area: 850 m²
主要材料：白沙米黄大理石、虎檀尼斯、泰柚	Major Materials: White Sand Beige Marble, Sandalwood, Thai pomelo
摄影：江河摄影	Photography: Jianghe Photography

本项目以柏悦酒店为依托，傍依浙江宁波东钱湖自然景区，独享小普陀、南宋石刻群等人文景观资源，地理位置无可比拟。在空间和视觉语言上与柏悦酒店完美对接。在空间上以中国建筑传统的空间序列强化东方式的礼仪感和尊贵感，在视觉上通过考究的材料和独具匠心的工艺细节，以简约的黑白搭配一气呵成，展现了东钱湖烟雨濛濛、水墨沁染的气韵。

在硬件和智能化体系上坚持柏悦酒店一贯高品质的传承，让客户不经意间感受到骨子里的柏悦性格。比如：一进入会所，所有的窗帘为你徐徐打开，让阳光一寸寸地洒进室内；按一下开关，卫生间的门就会自动藏入墙内；全智能马桶自动感应工作……随处都力求让人感受到高品质的舒适体验。

设置独立专属的高端客户接待空间，如独立酒水吧、独立卫生间。尽享尊贵、专属的接待服务。细分功能空间，将一个空间的多重功能拆解细分，每个都尽善尽美，大大提升品质感。

增加全新的功能体验，在商业行为中加入文化和艺术气质。我们在地下一层设计了一座小型私人收藏博物馆，涉猎瓷器、玉器、家具、中国现代绘画等……不仅大大提升品质，同时也给客户带来视觉和心理上的全新震撼体验。

身处其中，恍若超脱凡尘，烦恼、杂念消失无踪。带出一抹我独我乐的欢喜。

This project is set off by Park Hyatt and is beside Dongqian Lake Scenic Spot of Ningbo city, enjoying the human landscape resources of Little Putuo, carved stones from the Southern Song Dynasty, etc. Thus this club has incomparable geological locations. This club combines perfectly with Park Hyatt in space and visual languages. It strengthens oriental etiquette sensation and noble feelings with traditional Chinese architectural space layers. Visually this club has concise and smooth black and white collocations through exquisite materials and ingenious craft details, displaying misty charms with tradition Chinese painting style of Dongqian Lake.

As for the system of hardware and intellectualization, it sticks to the inheritance of the consistent high quality of Park Hyatt. Thus unconsciously the guests can perceive the essential Park Hyatt characteristics. For example, once you enter the club, all curtains would open slowly for you and sunshine would spread into the room inch by inch. Once you push the button, the door of the washroom would hide into the wall. All-intellectual closestool would carry out auto-induction work… All of these would assure people of high quality comfortable experiences.

This club has independent and exclusive high-end reception room, such as independent wine bar, independent washroom. All guests would enjoy honorable and exclusive reception services. All functional spaces were subdivided to disassemble the multi-functions of the space. The designer tries to attain the perfect status for every space to promote the quality in a grand scale.

The designer adds brand-new functional experiences, adding cultural and artistic temperament in the business activities. There is a small size library for private collections in the basement, including porcelain, furniture, modern Chinese paintings, jade ware, etc. All these not only uplift the quality, but also create brand-new shocking experiences for the guests visually and psychologically.

It is totally like another world. All worries and distracting thoughts would disappear. People would experience some exclusive joys here.

神韵 新中式空间设计典藏

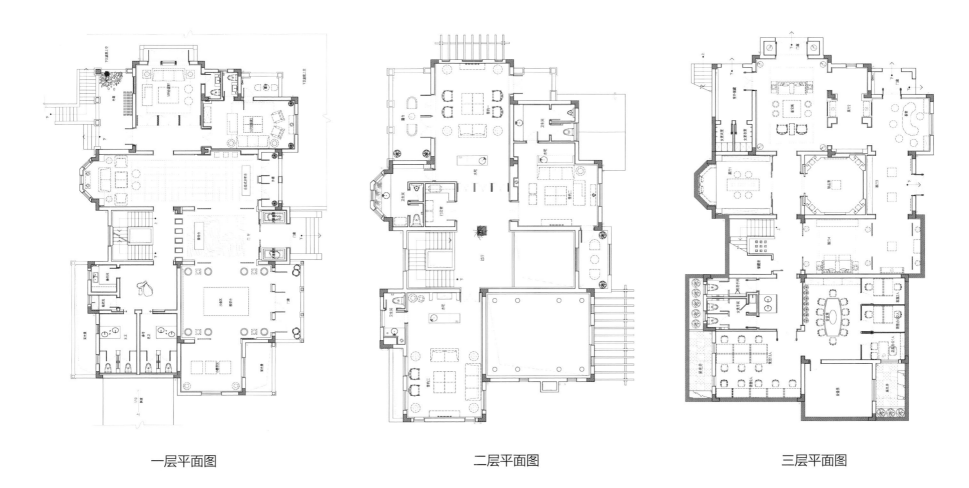

一层平面图　　　　　　　　　二层平面图　　　　　　　　　三层平面图

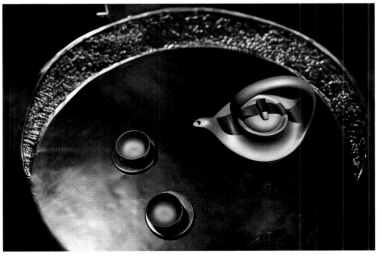

神韵 新中式空间设计典藏

The Forth Chapter
Dining Space

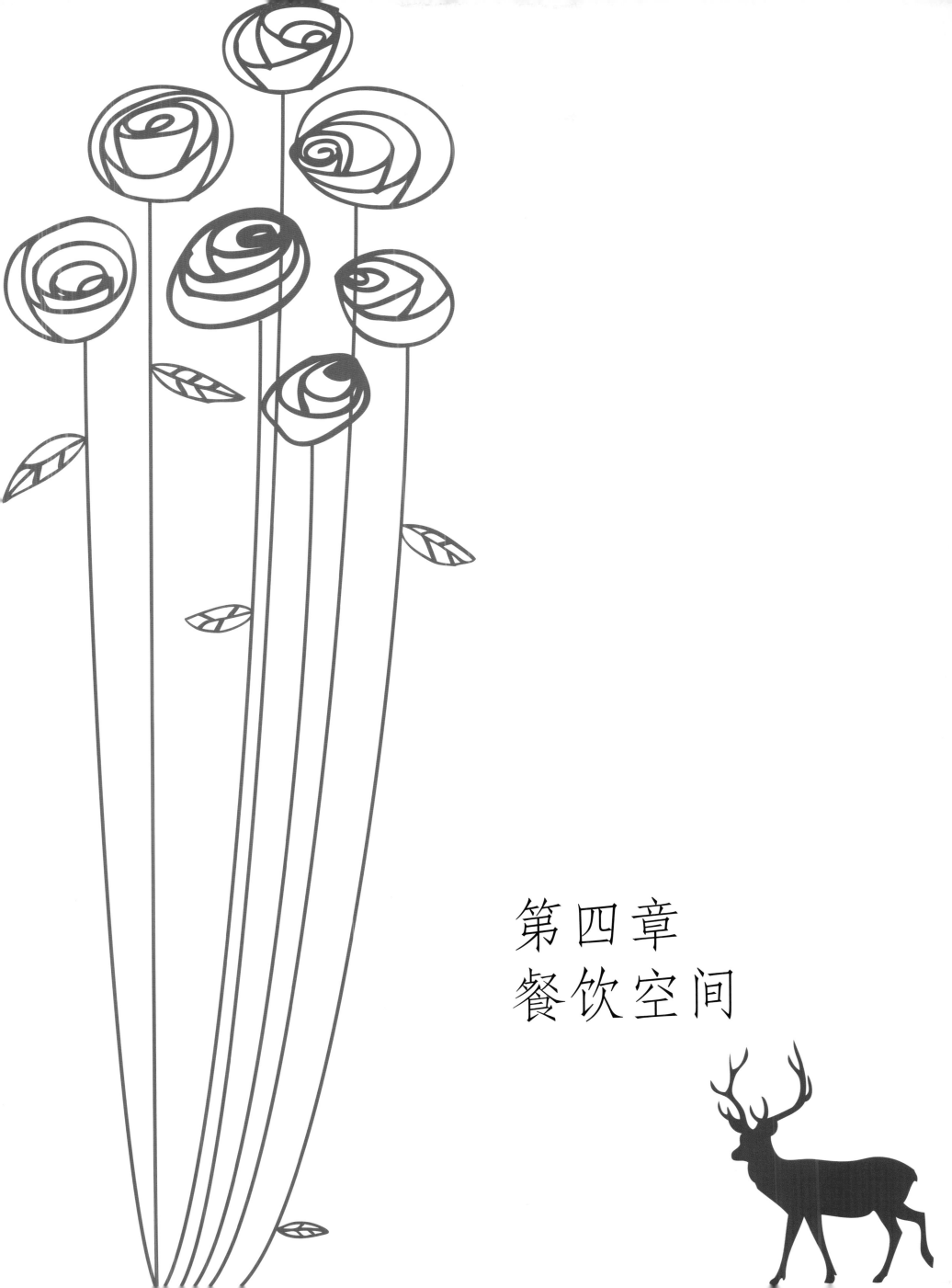

第四章
餐饮空间

水墨哲学
眉州东坡酒楼亦庄店
Ink Style Philosophy
Meizhou Dongpo Restaurant Yizhuang Shop

设计单位：经典国际设计机构（亚洲）有限公司	Design Company: Classic International Design Agencies (Asia) Limited
设计师：王砚晨、李向宁	Designer: Wang Yanchen, Li Xiangning
项目地址：北京亦庄经济开发区	Project Location: Beijing Yizhung Economic Development Zone
项目面积：1700 ㎡	Project Area: 1700 ㎡
主要材料：棕云石、银白龙、真丝壁布、印刷丝绸玻璃、激光切割铁板屏风	Main Materials: brown marble, silver dragon, pure silk wall cloth, silk printed glass, laser cutting iron screens

设计师运用现代的手法，演绎中国传统文化的内在精神，展示新材料和传统材料的装饰效果。利用水墨方式呈现中国传统哲学的处事之道，使空间充满灵动的自然之美。

通过传统材料的运用，唤起人们的想象，来取得意想不到的效果。空间中运用大量的传统丝绸面料，与新的玻璃材质相结合，创造出充满自然美又兼顾时代感的材料，优雅地表现传统艺术的温婉雅致。

在四层电梯的入口处，一面仿古铜镜映衬出的画面着实令人叹为观止，对面的铁艺雕刻屏风巧妙隔开等候休息区和收银区，传统的中式屏风，经最现代的钢铁材料和激光技术的重新演绎，呈现出迥然不同的审美情趣。

空间中大幅水墨山水壁画的载体不再是传统的宣纸和绢，而是丝绸玻璃。东坡泛舟赤壁的经典画面也被重新解构，山水和人物被分为两层，在玻璃和丝绸的映衬下，随着视线的移动，山水和人物在空间中形成新的视觉映像。这些层次变化的界面将内部空间进行了重新定义，顺应了空间的美学需要，并传递了某种动感的旋律，使人不再感到置身在一个静态的空间中。

Designers try to use modern techniques to deduce the inner spiritual essence of traditional Chinese culture, to show the decorative effect of new materials and traditional materials. Using forms of Ink to present the way dealing with affairs of traditional philosophy, which made space filled with natural beauty.

Through the using of traditional materials, the design arouses people's imagination and gets unexpected results. A large number of traditional silk fabrics are used in space, and combined with new glass to create a material fulling of natural beauty and period feeling and gracefully express gentility and elegant of traditional art.

At the entrance to the fourth floor of the elevator, the picture silhouetted by an archaistic bronze mirror is really amazing. Wrought iron sculpture screen in the other side separated the waiting lounge from the cashier area artfully. Through the most modern steel materials and laser technology, traditional Chinese screens have been reinterpreted and present a very different aesthetic taste.

The carrier of large ink landscape paintings in space is no longer the traditional rice paper and Juan, but silk glass. Classic picture of Dongpo white water rafting Chibi boating has also been re-deconstructed. Against the background of glass and silk, with the moving of sight, landscapes and characters which have been divided into two layers form a new visual image in space. These changed interfaces has redefined the interior space, complied with aesthetic needs of the spatial span and passes some kind of subtle movement, which makes people no longer feel in the middle of a static space.

神韵 新中式空间设计典藏

神韵 新中式空间设计典藏

好客山东
郑州大风餐饮店

Friendly Shandong
Zhengzhou Da Feng Restaurant

设计公司：河南东森装饰工程有限公司	Design Company: He Nan Dong Sen Decoration Engineering Co., Ltd.
设计师：刘燃	Designer: Liu Ran
项目地点：郑州	Project Location: Zhengzhou
项目面积：600 ㎡	Project Area: 600 m²
主要材料：青砖、汉瓦、鸟笼、红灯笼、中式花格	Main Materials: Black brick, han tile, birdcage, red lantern, Chinese lattice

本案位于中原之都——郑州，建筑面积600㎡。整体方案以山东文化做背景，容纳山东各地的风情元素。"借景望景、步移景动、曲径通幽"等设计内涵在这里得以充分体现。大堂设计着重体现大气内敛的气氛。"红灯笼"，烘托了整个大堂红火的氛围。包间设计以山东各地名胜为名，如"济南泉城、五岳泰山、孔圣曲阜、海滨青岛、菏泽牡丹"等，突出设计的主题。另外，在空间设计上留下了"室内建筑"的影子，青砖、汉瓦、白墙、鸟笼、红灯笼、中式花格等现代的设计手法贯穿于整个空间中，使人瞬间就会被这热情似火的氛围所感染。

This project is located in the capital of the center plain-Zhengzhou, covering an area of 600 m². The design is based on Shandong's culture and local custom. It well represents the Chinese style that "walk to see more, each step brings you different scenery and a winding path leads to a secluded place." Red lanterns in the lobby contribute to the warm phenomenon. The names of separated rooms adopt that of famous places in Shan Dong, such as "Spring City Ji'nan" "Mountain Taishan" "Heze Peony". Different elements are arranged with modern techniques, embodying the warmth and enthusiasm of this restaurant.

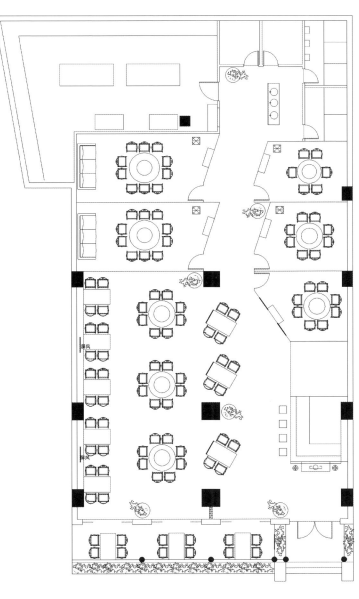

平面布置图

桥亭回廊间旧梦重温
桥亭活鱼小镇

Renew Old Romance in the Bridge Pavilion Gallery
Bridge Pavilion Fish Town

设计单位：福建东道装饰设计　　Design Company: Fujian Host Road Decorative Design
设计师：李川道　　　　　　　　Designer: Li Chuandao

传说"桥亭"源自一个溪多、桥多、亭多的桥亭村，那里的村民好以鱼待客，烹煮出的鱼别具风味，具有淳朴的味道。本案设计师结合该品牌的文化内涵，秉承其一贯的仿古风格，使设计尽显雅致韵味。

青砖石板旧廊桥，不过两百余平米的面积内，设计师将记忆里的老画面一帧一帧的回溯，这个迎来送往的商铺以开门见山的方式接客，原汁原味的旧木门敞开着。进门走道的两边，一半是前台，一半是入口。入口待客区是廊桥上标志性的座椅，俩小儿荡秋千的童真童趣的场景、都市里不常见的扎染粗布的场景，怕是再急着进餐的宾客也愿意多看几分钟。

天然的石磨、黑白的老挂画、灰白的绒布软垫，连搭建的木材都是褪色的，像是经过风雨飘摇的桥亭。它虽然失去了原本光鲜亮丽的色彩，却多了一番值得反复寻味的情愫。与大堂的古朴老旧色彩相比，回廊里的景致更为华丽。石墩和大圆柱是乡村里必不可少的元素，大红灯笼高高挂，像是节日里的张灯结彩，热闹非凡。旧时趴在圆木上与小伙伴嬉戏打闹奔走的画面历历在目。

复古色彩浓郁的餐厅里，品味的不仅是大鱼一条小菜三碟，还有"记得当时年纪小，你爱谈天我爱笑"细腻情感。设计师呈现的也不再是单纯的餐饮空间，像是造梦者，带着宾客在桥亭回廊间重温旧梦。

"Qiaoting", the story goes, is derived from a bridge pavilion village with many streams, bridges and pavilions, where the villagers treat the guests with fish which is cooked with unique flavor and honest taste. The designer combines cultural connotation of the brand and adheres to antique style to make the design show elegant charm.

Blue bricks, flagstone, old covered bridges. Within the two hundred meters area, the designer recalls the old picture frame by frame. This welcoming visitors and seeing them off shop greets guests in the form of coming straight to the point with the authentic old wooden door opening. One side of the entrance door aisle is the front, the other side is the entrance. Entrance hospitality area is the symbolic seat on the covered bridges. Even guests who are hurry to dine are willing to look the childlike innocence scene of two children swinging and the city's unusual scene of tie-dyed denim for a few more minutes.

Natural stone, black and white old paintings, gray velvet cushions, and even the built timber are fading as storm-beaten bridge pavilion. Although lost its original bright color, it has more repeated savor feelings. Compared with the quaint old colors of the lobby, corridors' scenery is more gorgeous. Stone piers and large cylinders are the essential element of villages. the Red Lanterns are highly hung like lanterns and festoons in the festival, lively and extraordinary. The picture of once lying on logs and frolicking with small partners is visible before the eyes vividly.

In the rich vintage color Restaurant, not only can we taste a big fish and three dishes, but also delicate emotions that "when we are young, you love talk and I love to laugh". Designer is no longer to present a simple dining space, like a dream-maker, but brings guests back into the dream in the bridge pavilion corridors.

神韵 新中式空间设计典藏

大宋情怀
工三便宜坊

Song Dynasty's feelings
Gongsan Bianyifang

设计团队：和合堂设计团队	Design Team: He Hetang Design Team
设计师：王奕文	Designer: Wang Yiwen
项目地点：北京工体北路	Project Location: Gongti North Road, Beijing
项目面积：2100 ㎡	Project Area: 2100 ㎡
主要材料：石材、真石漆、金色特殊漆、装饰灯具、绘画作品、印纱画	Main Materials: Stone, Lacquer, Gold Special Paint, Decorative Lamps, Paintings, Printed Yarn Paintings
摄影师：孙翔宇	Photographer: Sun Xiangyu
撰文：王奕文	Writer: Wang Yiwen

设计师赋予此空间"大宋情怀"的主题，以宋朝山水、花鸟画作、词牌意境等为载体的"叙事"方式来演绎空间的博大气势。

高高在上的亭台，将空间分为两个区域，一侧为挑高达9m、气势磅礴的婚宴区，一侧为传统意义上散座区。宋代山水画，博大如鸿，飘渺如仙，意境挥洒如行云，"亭台"宛若山水画之中的一处雅居，水波荡漾，树影婆娑，鸟语花香，将人们带入幽雅的意境之中。

宋式变异回廊的呈现某种意义上界定了空间的延续性，作为主要的动线，承载着重要的功能。两边配以曼妙的灰色轻纱，演绎出"无意苦争春，一任群芳妒"的姿态。两个大包间满足了高端商务的需求，宋代仕女的服饰，严谨的家具配色，工笔花鸟的绘画作品，演绎出春深雨过西湖好、百卉争妍、蝶乱蜂喧的动人景象。

The designer endowed the space with the theme – "Great Song Feelings". He adopted "Narration" way that as the Song Dynasty landscape, flower and bird paintings and the tune of artistic conception for materials deduce great of space.

The space was divided into two parts by the high above pavilion. There is a reception area that nine meters high and of great momentum on one side and the traditional casual sitting areas on another side. Song Dynasty landscape painting: great like Hong, ethereal like fairy, the artistic conception beauty like clouds. "Pavilions" like a garden in landscape painting, rippling water, shady trees, singing birds and fragrant flowers, all of these landscape can bring people into a quiet and tastefully laid out atmosphere.

To some sense, the presentation of Song type variation corridor defines the continuity of space, as the main routes, carrying important functions. Both sides with graceful grey veil deduces a kind of attitude that he doesn't intend to contest for the glory of Spring but he would rather be alone and envied by other excellent people. Two large rooms meet the high-end business demand. The Song ladies clothing, rigorous harmonize colors furniture and meticulous flower and bird paintings deduces a charming spectacle that like the West Lake after rain, flowers competes in splendor and bees and butterflies rise and dance in a happy mood.

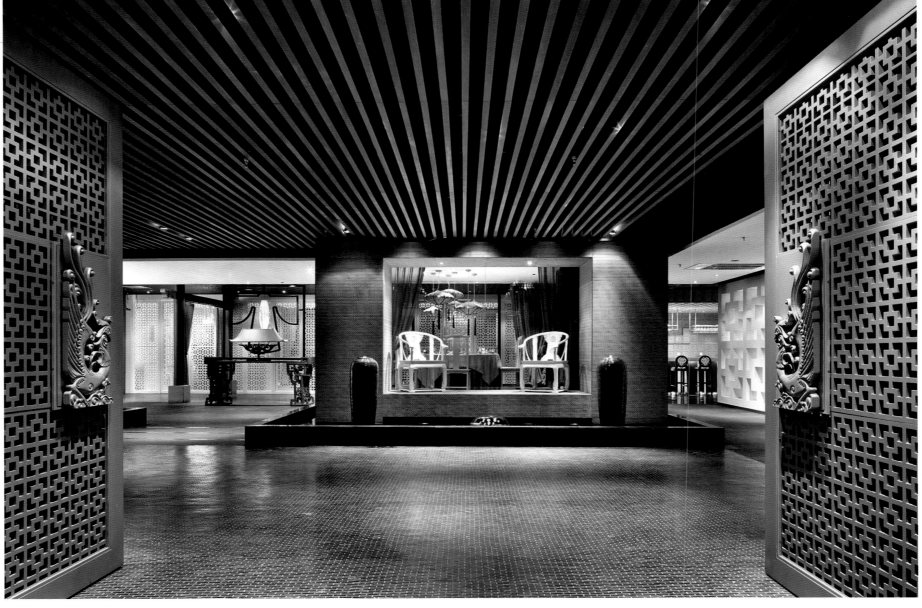

神韵 新中式空间设计典藏

平面布置图

神韵 新中式空间设计典藏

浪漫樱花
赤坂日本料理
Romantic Cherry Blossoms
Akasaka Japanese Cuisine

设计单位：品川设计顾问有限公司	Design Company: Shinagawa Design Consultants Limited
设计师：郭继	Designer: Guo Ji
项目面积：350 ㎡	Project Area: 350 ㎡
主要材料：黑钛、水曲柳木饰面、砂岩、瓷砖、金刚板、青石	Main Materials: black titanium, Manchurian ash wood veneer, sandstone, ceramic tile, stone, diamond plate
摄影师：吴永长	Photographer: Wu Yongchang
撰文：Makye	Writer: Makye

在日式风格的设计理念中，设计师一般比较注重体现浓郁的日本民族特色，在选料上注重质感的自然和舒适，常选择木格拉门、地台等元素来表现。在本案设计中，设计师除了保持日式餐厅的传统风格外，更增添了一些现代时尚的符号。餐厅内部环境雅致且轻松简约，布局独具匠心，散发着东方古朴的异域风情。主要材料为日式设计中常见的木质材料，将不同颜色质感的木制相融合，配合砂岩、金刚板、青石等比较刚毅的元素，其刚柔并济的视觉效果为我们带来幽雅舒适的就餐环境。在布局上，设计师运用完美的衔接，使得每个区域都功能明确，却没有硬性的区分界限，一切显得那么自然融洽，吸引着人们去细细探究。

惊艳不需要绝对的夸张，细节的完备才可以成就个性的完美。在这个 350 ㎡的空间里，设计将时尚与传统的符号相互融合，巧妙地运用到每一处细节中，趣味和创意如舞动的精灵般吸引着人们的眼球。餐厅里随处可见的浪漫樱花，精致考究的的吊顶与吊灯，流淌着灵动气息的水池和鹅卵石，带有浓郁日式风格的屏风和拉门，还有那些造型优美且高雅舒适的餐椅，都是餐厅最为吸引人的亮点，体现了设计师的独具匠心。

无论是身处时尚典雅的大厅、还是围坐在日式风情的榻榻米包房里，置身在这样一个曼妙的环境中，感受它动人的情调，略显神秘的气氛，在这个轻松流动的空间，感受着一种高品质的生活，那是一种格调，同样也是一种人生。

Design concept in the Japanese style, designers generally pay more attention to reflect the rich characteristics of Japanese nation, in the choice of materials on the texture of natural and comfortable, often choose thewooden door, floor and other elements to the performance characteristics.In the case of design, designers in addition to maintaining the traditional style of Japanese restaurant, to add some more modern fashion symbol.The interior environment elegant and easy and simple layout, have great originality, the distribution of the ancient exotic oriental. The main material for common wood materials in Japanese, combining the different colortexture of wood, with sandstone, Jin Gang plate, stone and other relativelystrong elements, the softness of the visual effects bring us comfortable and elegant dining environment. In the layout, the designers use the cohesion and technique perfect, make each area clear function, but no hard and fastdistinction, everything seemed so natural and harmonious, attracts people to explore in detail.

Amazing doesn't need to be absolutely exaggeration, perfect perfect detailscan achievement of personality. In this 350m² space, designer fashion andthe traditional symbol of mutual integration, skillfully applied to each link, fun and creative as the dancing elf attract people's attention. The restaurantcan be seen everywhere in the romantic cherry blossoms, elaborate ceiling and pendant, flowing pools and pebbles flavor smart, screen and the doorwith a strong Japanese style, the elegant and graceful and comfortablechairs, are the most attractive spot for the restaurant, the designers have great originality.

Whether it is in fashion and elegant lobby, or sitting in the Japanese styletatami rooms, in such a graceful environment, touching sentiment,somewhat mysterious atmosphere, the flow in the comfortable space,experiencing a high quality of life, it is a kind of style, is also a kind of life.

神韵 新中式空间设计典藏

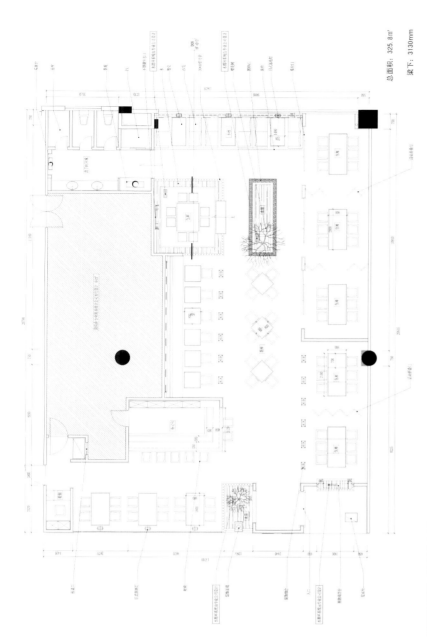

平面布置图

神韵 新中式空间设计典藏

春天来了
春天里新川式健康火锅
Spring is Coming
Chuntianli New Sichuan Health Hot Pot

设计公司：成都龙境品牌策划有限公司	Design Company: Chengdu Long Jing Brand Planning Co., Ltd.
设计师：毛继军	Designer: Mao Jijun
项目面积：1400 ㎡	Project Area: 1400 ㎡
项目地址：成都	Project Location: Chengdu City

记得一位台湾建筑师说：建筑的本质是一种陪伴。什么是陪伴？我理解就是让空间和你一同慢慢老去。空间，一定要能给人温暖，让人亲近。重新发现藏在日常生活中的美，回到田园，自然成了本案的立意之源。

"春天里，桃花开"，桃红柳绿中，田边农舍伴着炊烟，夕阳的光穿过木窗溜进了灶台……这一幅田里乡间的景象，构成了本案所要表达的空间情怀，顶部正如梯田，一块块的并置着，田边的小道正好拿来将灯藏入，只留下光。地面如屋后面的竹林，一条条地错落着，和着泥土的颜色，仿佛能听见春风的声音……

日本设计师原研哉说："只有空的容器，才有装入无限东西的可能。"

的确，空纳万境，间中有情。

I remember that a Taiwanese architect said: The nature of the architecture is a kind of companionship. What is companionship? My understanding is to make space and you grow old together slowly. There is a feeling in this point of view. Space–it must give people warm and let people close. Rediscover the beauty hidden in everyday life and return to the countryside, so nature became conception of the case. "

In spring , peach blossoms ," in the pink peachblossoms and green willows. Farmhouse in the farmland accompanied by smoke, sunset light through the wooden windows slipped into the stove ... this image of field and countryside, which forms space feelings that this case want to express. Roof just like terraced fields placed in rows. The trail in the farmland can hide the lamp, leaving only light. The ground just like bamboo grove behind the house scattered in rows. It follows with the color of mud and it seems that we are hearing the sound of the wind.

Japanese designer Kenya Hara said:" Only empty containers can it has the possibility that can place infinite things."

Indeed, empty can contain infinite things and embody feelings.

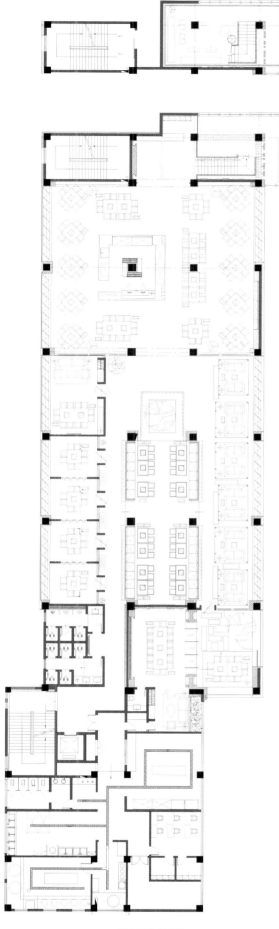

平面布置图

锦绣中国红
福州海通一号（梅峰店）
Fairview Chinese Red
Fuzhou Haitong One (Meifeng Shop)

设计公司：福州国广装饰设计工程有限公司
设计师：丁培瑞

Design Company: Fuzhou Guo Guang Decoration Design Engineering Co., Ltd.
Designer: Ding Peirui

设计师利用水面的效果和光影的微妙变化创造出虚实互补的感觉，潜移默化地影响着各个区域的气质，带来沉静、内敛的气氛。中式风格的包厢则以另一种姿态呈现，将现代手法与传统模式有机结合，让空间充满东方的审美韵味。

在略显空旷的公共大厅，设计师运用"加"法对空间进行规划：一组充满水墨意象的大型吊灯、一尊神秘绚烂的木制雕塑、一幅传统山水画、一面辅以荷叶装饰的背景墙，还有加建于大厅正面、贯穿两层高的楼梯，这些场景实现着空间外在和内在精神的展示。

包厢内部在沿袭中式气质的基础上，通过实木和石材对比，丰富了空间层次，加强了餐厅的现代气息。作为集餐饮、休闲为一体的大包厢，透过层次的构建，由内而外、由紧密到舒缓，环状层次的交叠衍生出具有雕塑感的空间氛围。

包厢的内部装饰采取中国传统的古典元素——"回"形木栅格、扶手椅、罗汉床、水墨画，还有极具东方意味的中国红坐垫。以传统元素作为贯穿始终的语言，增添了人文气质。与此同时，还保留了传统经典的明代家具元素，在材质、纹路、造型精益求精的基础上，加之石材的硬朗及磅礴之气，为中式古典空间注入了简洁利落的气势。

Designer uses the effect of water and subtle changes in light to create virtual complementary feeling. It affects temperament of every region inconsciently and bringing quiet, restrained atmosphere. Chinese style box presents in another way and combined with modern techniques with traditional pattern, so that space is full of oriental charm of beauty.

In the slightly open public hall, designers use the "plus" method to plan space: A set of full of ink image chandeliers, a mysterious and gorgeous wooden sculpture and a traditional landscape painting and a Lotus leaf decoration background wall. Besides, the staircase that was built in the front of the hall and connected two floors, these scenes achieved external display and inner spirit of the space.

Interior of boxes based on the following Chinese temperament and contrasting the wood and stone has enriched the space level and strengthened the modern of the restaurant. As big boxes that set dining and leisure as a whole, through the hierarchy construction, from the inside to out, from the close to the soothing and annular level overlap derived the sculptural of space atmosphere.

Interior decoration boxes adopted the traditional Chinese classical elements – the "Hui Zi" –shaped wooden grid, armchairs, Lohan bed, Chinese monochromes and very oriental Chinese red cushion. Adopting the traditional elements as the throughout language brought more humanistic temperament. At the same time, it retains the traditional classic Ming Dynasty furniture' elements and on the basis of material, lines and modelling excellence , combined with tough and majestic stone then it pours simple and neat feeling into Chinese classical style.

神韵 新中式空间设计典藏

神韵 新中式空间设计典藏

黄河谣
中华国宴

The Yellow River Ballad
The Chinese State Banquet

设计公司：柏盛国际设计顾问有限公司	Design Company: Bai Sheng International Design and Consulting Co., Ltd.
设计师：杨彬	Designer: Yang Bing
项目地址：郑州市郑东新区CBD步行街	Project Location: Pedestrian Street, Zheng Dong New District CBD, Zhengzhou City
项目面积：6000 ㎡	Project Area: 6000 ㎡
主要材料：青砖、青石、木雕、砖雕、国产黑金花大理石、樱桃木饰面、象牙白乳胶漆	Main Materials: Brick, bluestone, wood, brick, domestic black gold flower marble, cherry wood finishes, ivory paint

"中华国宴"遵循传统文化的脉络，体现中原文化的精髓，将时代感、历史感都融入于空间的气息里。设计师以"在室内营造园林"的设计思路，将6000㎡空间的大气磅礴与室外的园林景观和建筑结合起来，以新东方主义为基调，表达出淋漓尽致的东方美，而与之相呼应的黄河文化也在空间中占有一席之地。设计师将当地传统工艺进行再创造，成为时尚装饰艺术的典范。同时在精品柜及墙面上陈列着极具河南特色的艺术品，运用河南历代名家诗词点缀空间主题，使空间闪耀着璀璨的光辉，诉说着"中华开国第一宴"的主题故事。

"The Chinese State Banquet" follows the context of traditional culture, reflecting the essence of the Central Plains culture. Integrate the sense of the times and the sense of history into space atmosphere. Designer adopted "Indoor construction landscape" design expression to connect the grand and magnificent 6000 square meters of space and landscape and building that outdoor together, and adopted new orientalism as main key, so express the most incisive oriental beauty, and the Yellow River culture with echoes of having a share in the space. Designers recreated the local traditional crafts and become a model of fashion decorative art. At the same time, they display very characteristics of Henan Art on the boutique cabinets and the metope. Adopting the Henan masters poems to embellish the theme, so that the decorative space shine with dazzling brilliance and telling "the founding the first feast" the story of the theme.

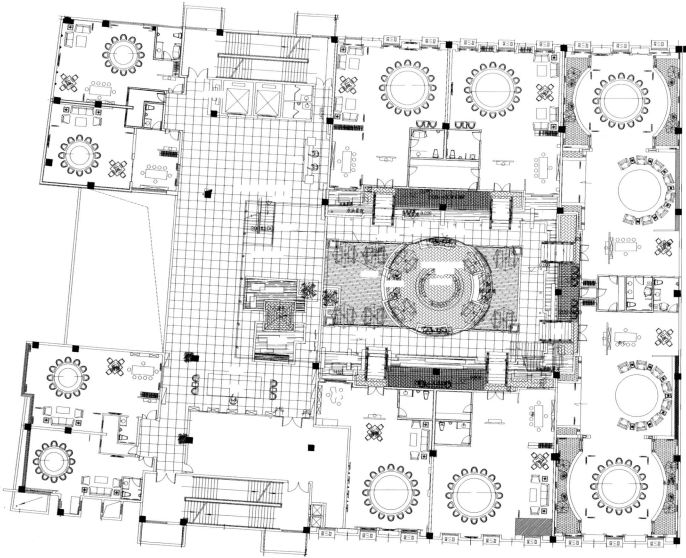

平面布置图

苍劲中的柔和美
渔人码头时尚鱼火锅
Vigorous in the Soft Beauty
Fisherman's Wharf fashion fish Hot pot

设计单位：楼语设计工作室	Design Company: Lou Yu Design Office
设计师：林金华	Designer: Lin Jinhua
项目面积：680 ㎡	Project Area: 680 ㎡
主要材料：仿锈砖、壁纸、造型藤灯、大理石	Main Materials: Fake rusty brick, wallpaper, bine lamp, marble

进入渔人码头的时尚火锅店中，藤灯与防锈砖、大理石与壁纸交糅在一起，既矛盾又和谐，有一种苍劲的美，又似乎若隐若现地展示出一缕淡定的情愫，给人视觉上的冲击。设计师在进门的大厅处放置了一个红色的、憨态可掬的人物雕塑，它既起到了迎客的作用，也与前台背景墙上的书法装饰形成刚柔对比，不着痕迹而又颇具用心地将传统气息沁入其间，即使是浮躁的心绪也会在不知不觉中安静了下来。

在用餐区，既有卡座，也有圆弧形的小包间。带来了奇妙而趣味十足的感官享受。在整体的暗环境中，这些光影的变化使空间的气息有了一丝流动的意味，仿若重复滚动播放的胶片上，闪烁着海市蜃楼的身影，给人似曾相识的体验，激发出一种愉悦的空间体验。

独立包厢中，设计师或以蓝色的背景装饰来营造水的气息，或以黄色的挂画带出温暖的质感。这些丰富的元素融入于周遭的环境中，丰富了空间的层次关系。

Sitting in this restaurant, you will feel a sense of harmony and contradictory through those lamps, marbles and wallpapers. It looks vigorous but also comfortable. The designer set a cute red figure sculpture at the entrance in the hall to send welcome to everyone comes. On the other hand, the sculpture becomes a sharp contrast to the calligraphy on the background wall, embodying the traditional elements without any notice, which helps customers calm down.

In the dining area, you can either choose the booth or private rooms. When the light goes dark, the air seems to move with the change of light streams. You shall know how pleasant that would be.

The designer chooses blue or yellow to decorate the private rooms. Those elements are comparative with environment and color the layer.

神韵 新中式空间设计典藏

空谷生幽兰
妙香素食馆
Orchard in the Valley
Miao Xiang Vegetarian Diet

设计公司：福州大木和石设计联合会馆	Design Company: Fu Zhou Da Mu He Shi Design Union Hall
设计师：陈杰	Designer: Chen Jie
项目地址：福州华侨新村	Project Location: Overseas Chinese Village, Fu Zhou
摄影师：周跃东	Photographer: Zhou Yuedong
撰文：全剑清	Writer: Quan Jianqing

素净淡雅的环境里没有繁杂的装饰，人们看到的是一份淡定和随性。造访"妙香素食"，更像前往信佛人家做客，品尝私房菜，追随他们一心向佛，却又不忘人间美味的那一份虔诚。

置身于妙香素食落英缤纷的院落里，呼吸着带有雨露泥土芳香的潮湿空气，远离了都市的喧嚣，让我们可以细细体会着这里的质朴和恬静。在清幽的环境中，主人将桌椅恰如其分地摆放在绿树丛中、走廊上，并用红伞、枯枝点缀其间，别有一番情趣。它们界定出的空间，用一种出人意料的美打动着每个走进这里的人。在绿树深处，假山流水之间忽见一间日式小屋坐落其间，屋顶上干枯的厚茅草，拉门与窗棂上的花格，屋内屋外古朴而灵动的陈设，都把我们带到了一种古老的意境里。

素食馆的室内保留了原始建筑两层小洋楼的格局，屋内主色调为白色，白色的墙、白色的门、白色的窗户、白色的屏风，一如素食带来的清新与纯净，让人忘却了都市的繁杂与欲望。大部分墙面上没有做过多花哨的装饰，只是在每个房间里恰到好处地挂上一些书法和国画作品，使得整个环境高雅起来。在楼梯的墙面上，设计师独具匠心地在上面手绘了一幅枯树的剪影，射灯在树梢间投射出一个暖黄的圆晕，如同一轮满月挂于枯枝之间，手法精巧、意境曼妙。

In a quiet and elegant environment, there is no complicated decoration. We shall feel a sense of calm and peace. To come here is like to visit Buddhists. We are appreciating the food as well as their loyalty to the Buddha.

Apart from the noisy world, this place provides us an opportunity to feel the simple and quiet life as seeing flowers blossoming and breathing the smell of soil. The tables and chairs are exquisitely setted in the shade of trees or on the aisle, which is decorated by red umbrellas and deadwood. They are attracting people in their unique way. Deep in the shade appears a Japanese-style cabin. Everyone will be led to an ancient mood by the thatch roof, the typical tracery and old displays indoor and outdoor.

Inside the restaurant maintains the original two-layer structure. The dominant hue is white-white wall, white door, white windows and white screen-like the vegetable that make people feel fresh and clean. On most walls hangs only a few calligraphy or traditional Chinese painting, which increase the elegance of the environment. On the wall beside stairs, the designer draws a sketch of a dead tree. In the treetops he sets a yellow halo, making the light like a moon hanging on the tree. This indeed is skillful and creative.

神韵 新中式空间设计典藏

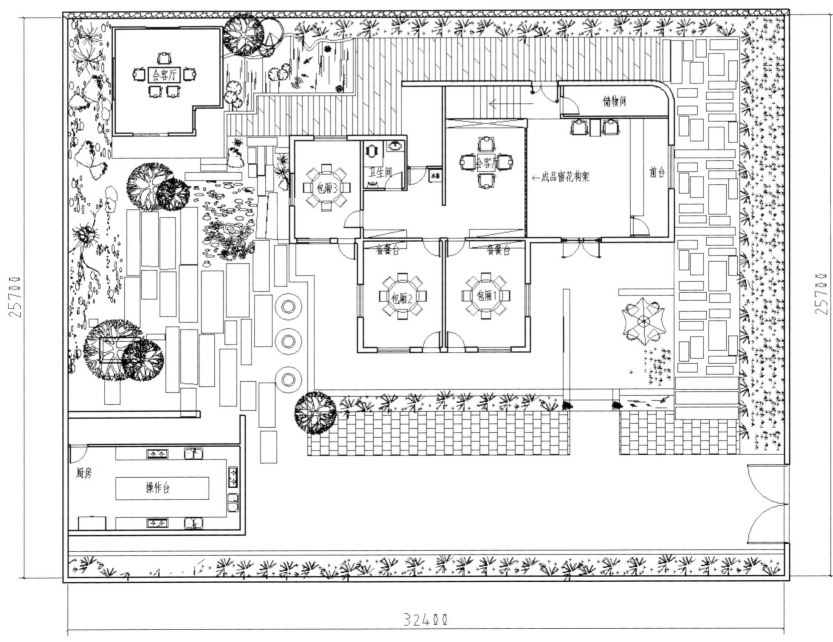

平面布置图

东方韵味
三义和酒楼
Ink painting of Dongqian Lake
A High-end Private Club in Yue Fu

设计公司：睿智匯设计	Design Company: Rui Zhihui Design
照明设计：睿智匯设计	Lighting Design: Rui Zhihui Design
设计师：王俊钦、许光华	Designers: Wang Junqin, Xu Guanghua
参与设计：陈丽娜	Participant: Chen Lina
项目地点：济南市历山路150号	Project Location: No. 150, Lishan Road, Jinan City
项目面积：2062㎡	Project Area: 2062 m²
主要材料：秋香木、肌理漆、彩绘、水磨石、金银箔、镜面不锈钢、皮革等	Main Materials: Qiuxiang Wood, emulsion paint, colored drawing, terrazzo, gold and silver foil, mirror finished stainless steel, leather, etc.
摄影师：孙翔宇	Photographer: Sun Xiangyu

三义和，一个极具东方韵味的知名企业，其前身为鲁西南风味酒楼。基于"打造中国鲁菜第一"的想法，为了谋求品牌化、高端化、国际化的发展思路，完成转型为高端餐饮的梦想，将旗下十余家鲁西南风味酒楼更名为"三义和酒楼"。

设计团队摒弃原有外观的复杂造型，因地制宜，整体呈现出一幅现代水墨画卷，并借用极具现代感的铁锈色框架材质搭配水墨画玻璃贴膜，用新现代主义手法表现东方韵味，使整体与济南千佛山的自然环境相互融合。

大厅运用麻绳、原木、瓷盘等材料，加入现代的彩绘手法，造成强烈的风格对比与视觉冲击，来彰显空间的个性。散座区主要运用"静中求动"的思想，色彩上，大胆运用中国传统的绿色及钛白色作为主色调。中心区的墙面和天花用山水书卷的形式，表达行云流水、延绵不断的效果。周遭以黑色钢制长管灯和叠级灯光，来增加空间的趣味性和韵律感。

VIP豪华包房和豪华大包房主要运用祥云和写意花鸟彩绘等元素，结合龟背纹、乳钉纹等中国传统的吉祥元素符号来打造。中小包房主要运用三义和企业的形象符号作为主元素，结合中式脸谱和宝箱花等中式吉祥元素为辅，结合充满现代质感的皮革和黑色镜钢，以极其现代的白色为主色调，将企业文化自然地融入整个空间中。

San Yihe is a well-known Chinese enterprise which is formerly named Southwestern Shandong Style Restaurant. The change of name aims at transferring those restaurants into high-end dining in a international direction and making the best Shandong cuisine in China.

The design team abandon the original complicated style, blending the whole building into the natural environment of Mount Qianfo. It looks like an ink painting which contains modern technique and traditional oriental features.

The decoration in the hall includes elements like hemp rope, log and porcelain cup, which are combined with modern colored drawing, causing a strong contrast in style and visual impact. In the seat area, designers insist on keeping a balance between movement and quiescence. They use green and white as the main color. On the wall and ceiling of the center part is a piece of Chinese landscape scroll, which shows a sense of free and simple. Around the scroll are different kinds of lights which can increase the rhythms of space.

The VIP private rooms and deluxe private rooms is decorated with auspicious clouds, colored drawing of birds and flowers and some other traditional lucky symbols. Middle and small private rooms is mostly decorated with the image of San Yihe and Chinese facial masks as well as other traditional elements. Those are blended with leather and black mirror finished stainless steel in white color, which well illustrates the company's culture.

神韵 新中式空间设计典藏

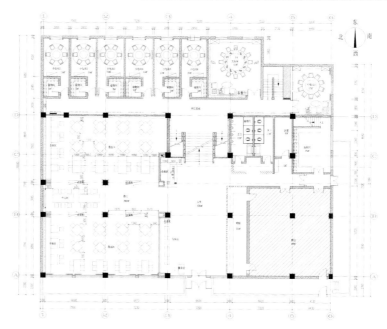

一楼平面布置图

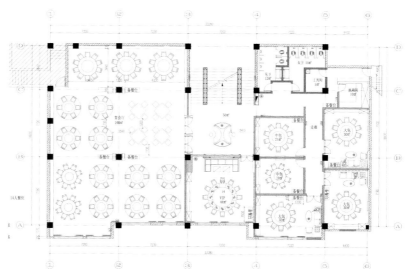

二楼平面布置图

神韵 新中式空间设计典藏

诠释四合院文化
唐会

Interpretation of Courtyard Culture
Tang Club

设计公司：上海唐玛国际空间设计有限公司	Design Company: Shanghai Tangma International Space Design Co., Ltd.
设计策划：施旭东	Design Strategist: Shi Xudong
设计师：洪斌、陈明晨、林明、胡建国、王家飞	Designers: Hong Bin, Chen Mingchen, Lin Ming, Hu Jianguo, Wang Jiafei
项目地点：福州	Project Location: Fuzhou
项目面积：300 ㎡	Project Area: 300 ㎡
主要材料：原木做旧、艺术漆定制、腐蚀耐候钢、亚麻、钢丝网	Main Materials: do-old wood, art custom paint, antiseptic weathering steel, flax, steel wire gauze
摄影师：周跃东	Photographer: Zhou Yuedong
撰文：全剑清	Writer: Quan Jianqing

在唐会中，走道区域的设计遵循传统文化的精神，叠石搭成的水幕墙，传递着潺潺的水声。白色的洗手台上方悬置一根管道，人走近时，水便能感应而下。这种人与水的交流，渗透出空间的人文气息。白色的地面与周遭的古朴斑驳形成视觉反差，诠释着阴阳协调的理念。

灰砖铺设的地面带有沉稳的力量，空间的吊顶及墙面用不同纹样的中式构件铺设，流动中展现出一种有秩序的大气之美。黑色在传统文化中代表"水"，而这种围合式的布局也潜移默化地诠释着四合院的建筑。摆设其间的家具以改良的中式设计为基调，用当下的使用习惯和审美需求反映出人们对传统文化的敬意。

二楼的区域中，三坊七巷建筑的写意形态与西方油画的写实静物搭配在一起，并在画的尽头虚拟上江水、帆船、海鸥等景致，让人不自觉地摇曳在艺术和文明的氤氲情境之中。另一侧的墙面上则轻描淡写着古代文人的形象，绵延其中的人文精神削弱了元素间的冲突，彼此的适度差异让空间充满了生动的气息。

Gray bricks laid ground reveals a steady power, ceiling and part of the wall in the space is carved with different patterns Chinese elements laid flows show a orderly atmosphere of beauty. Black represents the traditional culture, " water ", but also the layout of this Weiheshi subtle interpretation of the architectural practice courtyard. Furnishings furniture meantime Chinese design with improved tone, with the habits and needs of contemporary aesthetic reflects respect for traditional cultures. Stairs area, designers use modern geometric deconstruction collision and fusion of ideas to express a cultural and allow space between the light and heavy communication communion.

The ground was laid by gray bricks with a steady power, the ceiling and part of the wall was decorated by different carved patterns Chinese elements showed a kind of orderly beauty in the flowing.Black represents the "water" in the traditional culture, and this Weiheshi layout also interpreted the architectural practice of courtyard subtly. The furniture displays in it based on the improved Chinese style design and with the current habits that people had and aesthetic needs reflects people's respect for traditional culture.

The two floor area three square, Lane seven building freehand brushwork and the western oil painting still life together, and virtual River, sailing, etc. in the end of seagulls scenery painting, let people unconsciously swaying in the art and civilization dense context. On the other side of the wall is the ancient literati image touch on lightly, stretching the human spirit weaken the conflict between elements, moderate differences between let the space is full of vivid atmosphere.

神韵 新中式空间设计典藏

墨韵
平潭某餐饮会所
Ink Rhyme
A Dining Club in Pingtan

设计单位：上海唐玛空间设计有限公司	Design Company: Shanghai Tang Ma Space Design Co., Ltd.
设计师：施旭东	Designer: Shi Xudong
项目地点：福建福州平潭	Project Location: Pingtan Fuzhou Fujian
建筑面积：800 ㎡	Building Area: 800 ㎡
主要材料：木纹大理石、楼兰磁木系列、编织文硬包、	Main Materials: Wood grain marble, wood Loulan magnetic series, woven paper hard pack
摄影师：周跃东	Photographer: Zhou Yuedong

沿着楼梯婉转而上，错落有致且形态各异的金属吊顶、木色伞骨架和青砖石瓦般的四壁仿佛瞬间将人们带入了一幅东方命题的画卷。在这个充满想象的空间里，精神和意境、品质与灵魂、当代艺术和传统文化浪漫邂逅，一如设计师一贯的风格：强调新东方精神，强调艺术与空间的碰撞。通过残破的、被剥离的传统符号的抽象运用，寻找出最性感的地带，表达一种艺术的力量，并增强空间的神秘张力，用现代手法和工艺营造东方文化的艺术空间。

大堂中四个姿态不一的红人宛如一道屏风，设计师参考了唐朝《舞乐屏风》——以舞伎、乐伎为制作题材，锦袖红裳，人物飘逸俊美。"前音渺渺，笙箫笛筝，琵琶拍板，筚篥鼓叶"，这幅细说前朝场景的雕塑演绎了中国古代的贵族生活。于是生命在空间里充盈灵动，拥有了一份浪漫主义的气质。荷花在立体壁画和屏风中被大面积地凸显，"荷叶五寸荷花娇，贴波不碍画船摇。相到薰风四五月，也能遮却美人腰。"这出淤泥而不染的清廉之花深得古代文人雅士的喜爱，但从设计师手中表达出来，却有了一股婀娜多姿的自然之趣和大气深邃的东方意境。

Walking up the stairs along slowly, well-proportioned and diversed shapes metal ceiling, wood color umbrella skeleton and like brick tile walls seemed that suddenly bring people into an oriental proposition picture. In this full of imagination space, spirit and artistic conception, character and soul, contemporary art and traditional culture encounters romantically, as designers consistent style: emphasize the new oriental spirit and emphasize the collision between art and space. Through abstractly applying the sactered and stripped traditional symbols, they find out the sexiest region and express a kind of art power, so that can enhance the mysterious tension of space and use modern techniques and technology to create Oriental cultural art space.

In the lobby, the four different pose red statues like a screen. Designers refered to Tang Dynasty "Dance Music Screen" –used dancers, musicians as the production subject matter, kam cuff and red clothes, handsome and elegant characters. "Before the sound transitory man, farewell Zheng flute, lute finalized, wicker horn drum leaf ", this picture of the Yueji elaborate the former scene sculpture deduced the ancient Chinese aristocratic life. So in space, life filling Smart and it owns a romantic temperament. Lotus is highlighted by a large area in three-dimensional mural and screen." Lotus leaves grow into five-inch and lotus flowers delicate and charming, stickers wave does not hinder the master's shake. Grow to Xunfeng April and May, but also can cover up beauties' waist." This silt but don't dye incorruptness flower won the ancient literati's favorite, but from the expression of the designers' hands and there is a very pretty and charming natural interest and atmospheric profound oriental artistic conception.

神韵 新中式空间设计典藏

神韵 新中式空间设计典藏

舌尖上的岭南
舌尖岭南连锁餐厅
The Tongue of South of the Ridges
Tongue South of the Five Ridges Chain Restaurant

设计公司：深圳华空间设计顾问有限公司	Design Company: Shenzhen Hua Space Design Consultant Co., Ltd.
设计师：熊华阳	Designer: Xiong Huayang
项目地址：深圳	Project Project Location: Shenzhen
项目面积：480 ㎡	Project Area: 480 ㎡
主要材料：毛石、橡木、啡网大理石、玻璃磨花	Main Materials: Rubble, Oak, Emperador marble, Frosted glass

岭南餐厅是五谷芳新派特色餐饮品牌，以传承岭南文化为己任，集岭南的特色于一体，精研考究。餐厅设计采用现代时尚风格，咖啡色的大理石地面，毛石墙面，使餐厅有一种自然质朴的氛围。最有意思的是设计师将使用过的蚝壳清理干净，置于前台前，将特色美食以现代艺术的形式表现出来，这种设计创意得到了人们的高度认可。

South of the Five Ridges restaurant is the grains Fang New School South of the Five Ridges featured catering brand. Mainly responsible for the heritage of South of the Five Ridges culture and have the characteristics of South of the Five Ridges. It is finishing very well. Modern fashion style is adopted in restaurant design. Because of the coffee colored marble floor and rubble wall, there is a natural and plain ambiance in the restaurant atmosphere. The most interesting designers clean up the used oyster shells, put them in front of the reception desk and show feature of foods in modern art way. This design idea has been highly recognized by the market.

神韵 新中式空间设计典藏

神韵 新中式空间设计典藏

清歌一曲月如霜
台门酒会
Sing Like Cream
The Stage Door Wine Party

设计公司：杭州肯思装饰设计事务所	Design Company: Hangzhou Kent's Decoration Design Firm
设计师：林森、谢国兴	Designers: Lin Sen, Xie Guoxing
项目地点：杭州	Project Location: Hangzhou
项目面积：360 ㎡	Project Area: 360 ㎡
主要材料：青石、松木板、仿古实木地板、羊皮纸灯	Main Materials: Bluestone, Pine boards, Archaize hardwood, floors, Parchment lamp

项目位于杭州城皇山脚下，河坊街与南宋御街旁。河坊街，是一条有着悠久历史和深厚文化底蕴的古街，曾是古代都城——杭州的"皇城根儿"，更是南宋的文化中心和经贸中心。橙黄色的瓦片、青白色的骑墙、明光锃亮的牌楼，在冷暖相宜的光照下显得愈加韵味十足。

本案在结构上分为上、下两层，一层主要为散客品尝区、售卖区、文化展示区、总台区及操作间；二层则主要是VIP品鉴活动区。设计师在设计上，将老台门的精髓"陈、新、奢、纯"贯穿始终。

一楼主要采用新中式的手法，保留了绍兴老酒温婉的文化韵味，将酒、字画及自然的氛围融入其中。原建筑为木结构的房子，柱子较多，正是依靠这个特点，才有了柱林的创意。后来加入的柱子既作了阳光屋顶的支撑，又起到了区域连贯和文化展示的作用；蜿蜒的"小溪"横穿其中，让人联想到了古人饮酒作诗的场景，亦增加了品酒的情趣。

The project is located at the beside of Hefang Street and Southern Song imperial Street, the foot of Huang Shan mountain, Hangzhou. Hefang street is a long history and rich cultural foundation ancient street and was the ancient capital of Hangzhou "Beijing", moreover, it is the cultural center and trade center of the Southern Song Dynasty. Orange tiles, bluish white fence, glittering shiny archway, in the light of well-being seems increasingly affordable full of flavor.

In the case of the structure is divided into upper and lower two floors, the first floor is mainly for individual taste area, selling area, cultural display area, the total area and operating room; and the second floor is mainly VIP tasting area. Designers integrate the essence of the I "Old, new, luxury, pure" into the design consistently.

New Chinese style was mainly adopted in the first floor. They retained the Shaoxing wine gentle charm of culture and integrate wine, calligraphy and painting and natural atmosphere into it. There are more pillars in the original building that the structure is wooden structure houses, which is exactly to rely on this feature. This is why there is the later pillar of creativity. Later the joined pillars not only as the support of sun house roof but also played a role in the regional coherence and cultural exhibition; winding "stream" across it and reminiscent of the ancient drinking poetry scene and also increased the wine taste.

The Fifth Chapter
Offices Space

第五章
办公空间

禅意的平衡之美
郑树芬设计办公室
Zen Beauty of Balance
Simon Chong Design Office

设计公司：SCD(香港)郑树芬设计事务所	Design Company: Simon Chong Consultants Ltd.
设计师：郑树芬	Designer: Simon Chong
项目面积：250 ㎡	Project Area: 250 ㎡
项目地点：香港	Project Location : Hong Kong

郑树芬先生的每一个设计事务所办公室都风格一致，但又各具特色，有着"和而不同"的设计理念。郑树芬先生设计的作品无论是面积大小，都非常讲究"平衡"二字，比如"硬、软、暖"的相互结合。

在本案的设计中，软装的搭配是办公室的特色之一，郑先生喜欢旅游，去世界各地旅行时所收藏的艺术品，如古玩、字画、雕塑等摆放在空间中，琳琅满目却又和谐自然，不仅给办公区带来了轻松愉悦的感受，也同样是设计师创作灵感的来源。

Every design firm office of Mr. Zheng Shufen is with consistent and distinctive style, with " harmony in diversity " design philosophy. Whether the size of Mr. Zheng Shufen's design works is big or not, they are very particular about the "balance", such as mutual combination of "hard, soft, warm".

In the design of this case, the soft decoration is one of the characteristics of the Office. Zheng likes traveling. Works of art, such as antiques, paintings, sculptures etc., collected when Zheng travels around the world are placed in the space, a feast for eyes but harmonious and natural. It not only brings a relaxed feeling to the office, but also it is a inspiration source for designers.

神韵 新中式空间设计典藏

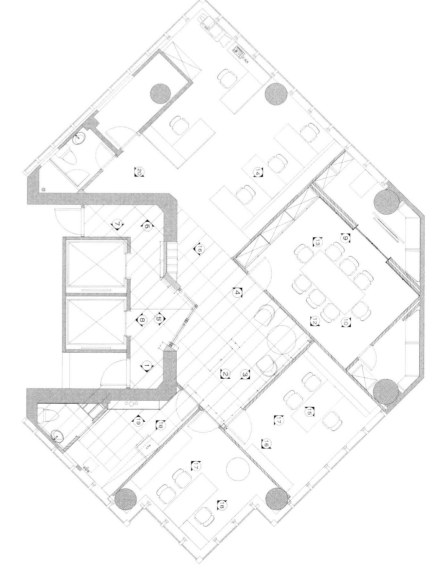

平面布置图

一行一世界 一静一禅心
无印良品空间设计办公室
A world of Quiet and Zen
Muji Design Office

设计单位：无印良品空间设计	Designer Company: Muji Space Design
设计师：陈邵良	Designer: Chen Shaoliang
项目地点：福建福州	Project Location: Fuzhou, Fujian
项目面积：700 m²	Project Area: 700 m²
主要材料：黑胡桃木塑、灰色硅藻泥、黑镜、素色水泥自流平漆、钢化玻璃、黑钛不锈钢	Main Materials: plastic black walnut, gray diatom mud, black mirror, plain cement self-leveling paint, tempered glass, black titanium stainless steel
摄影师：施凯	Photographer: Shi Kai
撰文：Zoe	Writer : Zoe

初入玄关，一目了然的中式意境就让人颇为惊喜。几簇新绿在大片的黑白之间，带来清新之感。开放式的空间，方正的格局、四平八稳的圈椅，简约凝练的线条。这些横平竖直的元素在不自觉间散发出中式特有的沉稳和泰然。

"坐亦禅，行亦禅，一花一世界，一叶一如来。"从等待区开始，这句耳熟能详的禅语便得到完美的诠释。在未熟悉地形之前，想要进入主体办公区也不是一件简单的事。看似通透近在咫尺的区域，却大有一番玄机。接待台的左、右两侧皆是透视的隔栅，呈严谨的对称式布局，中间则是圆形的镜面挖空。圆形的挖空通过空间的穿透成为自然的景框，让观看者将目光穿透至最里层的墙面上，古典园林造景的方式在钢筋水泥的办公室妙笔生花。

结束一番惊叹，推开侧边隐形成栅格的旋转门才算是真正进入办公区域。整个办公室拥有诸多的区域，风格上的求同存异则是设计上的巧思。光影旖旎，更是别具一格。大面积的白与黑形成对比，直接而不失委婉，简单而不减浓郁。高纯度的色彩带来最强烈、最有生气的视觉冲击。形态不一的格栅，将室内空间一分为二。一面是现代简约的工作台面，一面是古风犹存、气息淡然的空间布局。延续古色古香之意，木制镂空的旋转门在开合之间制造空间的变化，巧妙的设计中带着对中国文化底蕴的追寻。

Entering the entrance of Chinese artistic conception, stick out a mile to let people surprise. Several clusters of fresh green between the large black and white, to bring fresh feeling. The open space, the pattern of founder, chair stand stable, simple concise lines. The vertical and horizontal elements emitting a Chinese characteristic calm and serene unconsciously.

"Sitting is Zen, Zen line also, one world, a leaf of a Buddha." From the waiting area began this sentence for having heard it many times, usually can obtain perfect interpretation. Before the familiar terrain, want to enter the main office area also is not a simple matter. Circular hollowed out by spatial penetration is the natural view box, so that the viewer will look from penetrating to the innermost layer of the wall, the classical gardens in the ways of the reinforced concrete office write like an angel.

The end of a stunning, open the side contact into the revolving door grid is really into the Office Area. The office has a lot of areas, seek common ground while reserving differences on the style is the ingenuity design. Light is more charming, have a style of one's own. One is modern minimalist work table, one side is still the ancient atmosphere of space layout, indifferent, multi-faceted office space to let people surprise. The continuation of having an antique flavour of Italy, change the revolving door wooden hollow manufacturing space in between the opening and closing, clever design with beautiful pursuit of Chinese culture.

神韵 新中式空间设计典藏

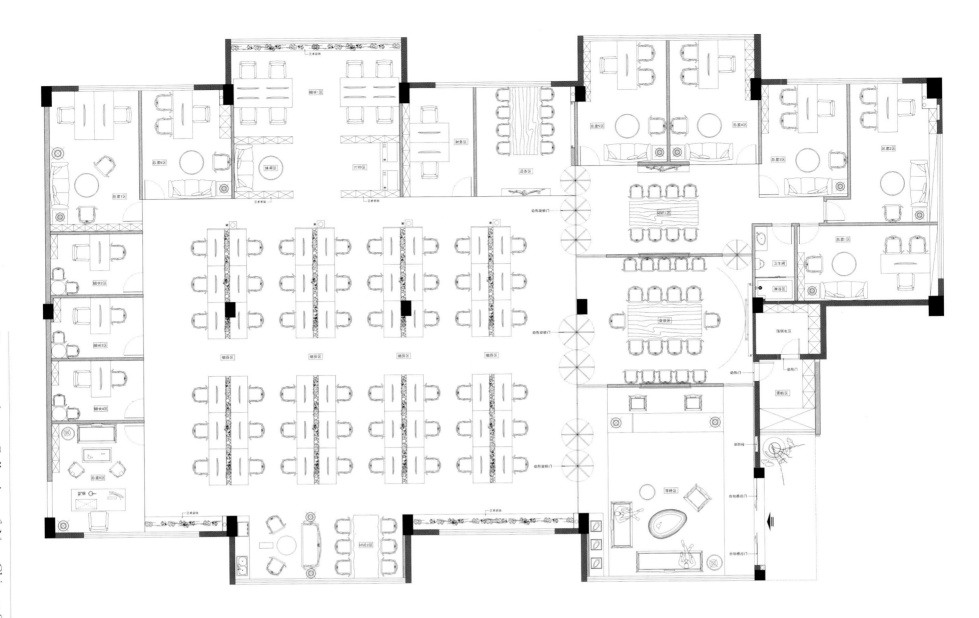

平面布置图

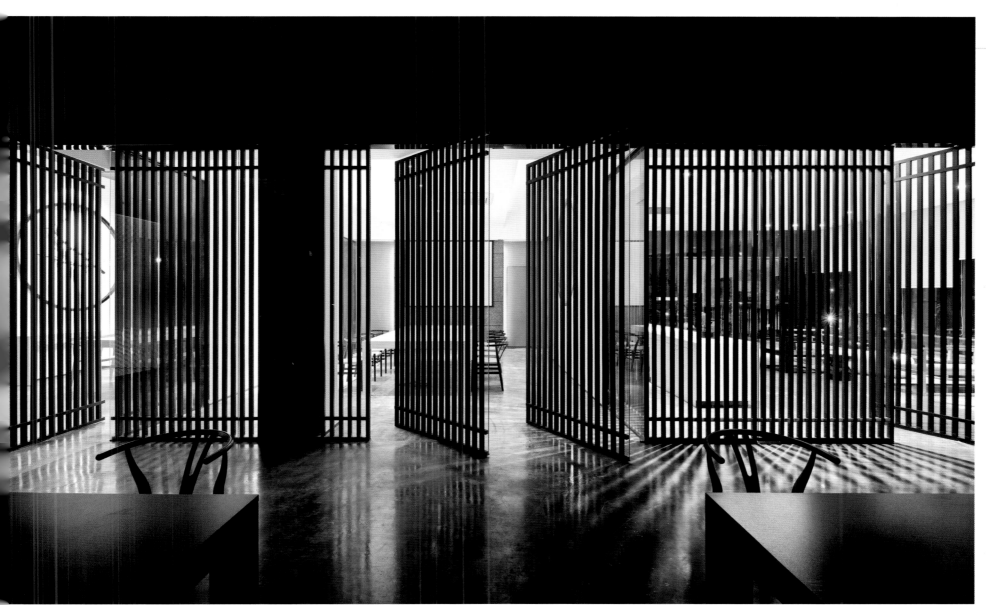

龙·鼎
某金融投资公司办公区设计
Dragon · Ding
Design of A Financial Investment Company Office Area

设计师：刘燃
项目地点：河南
项目面积：180 ㎡
主要材料：草织灯、树根、钢网

Designer: Liu Ran
Project Location: Henan
Project Area: 180 ㎡
Main Materials: Grass woven light, Shu Gen, steel net

本案为一家金融投资公司，设计师在其空间规划、设计手法及设计元素上扣准主题，采用低碳环保的材质（如：草织灯、树根、钢网等），打破了格式化办公空间的设计规则，体现出具有中国文化内涵的空间形式。

玄关处龙的符号以鼎的形式出现，并与中式条案及木龙雕形成一个有机的整体，成为本案的灵魂所在。悬吊的S灯（草织）与墙面悬吊的树根给洽谈区增添了轻松、自然的和谐氛围。会议室暴露式不规则的"金山块"，形成了天地呼应的视觉效果，并与公司性质相吻合。中式屏风成为空间的承载者，有机地把各功能区域紧密、合理的连在一起。

This case is a financial investment company. Designer Buckle quasi theme in its space planning, design methods and design elements, using low carbon environmental protection material (such as: grass woven light, Shu Gen, steel net etc.), broke the rule format design of office space, reflect the spatial form has Chinese culture connotation.

The entrance of the dragon symbol appears in the form of the tripod, and form an organic whole and the Chinese text and wood carving, become the soul of the case. S lamp suspended (woven grass) and the wall hanging roots to negotiate the area adds a harmonious atmosphere relaxed, natural. The conference room exposed irregular "Jinshan block", formed the world echo visual effect, consistent with the nature of the company and. Chinese screens become the space carrier, organic to each functional area, reasonable coherence together closely.

神韵 新中式空间设计典藏

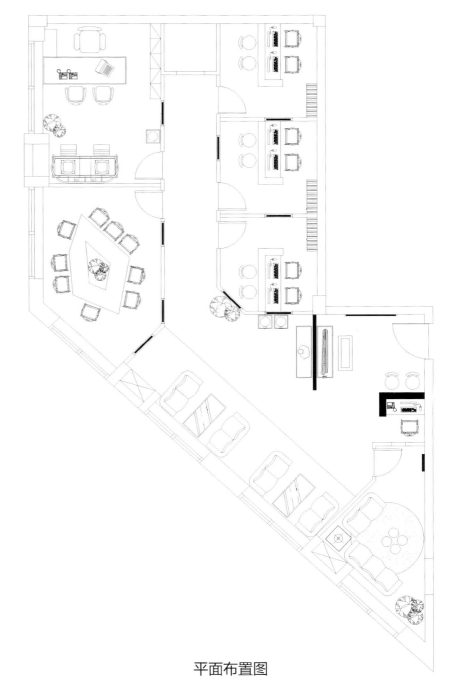

平面布置图

素色古味
设计师的工作室
Plain Ancient Flavor
The Designer's Studio

设计公司：合肥许建国建筑室内装饰设计有限公司	Design Company: Hefei Xu Jianguo Architectural Interior Design Co. Ltd
设计师：许建国	Designer: Xu Jianguo
项目地点：合肥	Project Location: Hefei
项目面积：130 ㎡	Project Area: 130 ㎡
主要材料：旧房梁、旧木板、旧窗户、红砖、乳胶漆	Main Materials: old, old wooden beams, the old windows, brick, latex paint

本案是设计师的工作室，设计时着重营造轻松自然的工作环境，寻找真正适合设计师自己需要的装饰效果，将轻松、随意的工作氛围引入室内设计中。

方案的设计注重以人为本，设计师将自己的喜好、思想、理念全部灌注在这个不过100多平方米的空间内。在这里，你既可以看到现代风格的简约家具，也可以看到古韵十足的明式家具，更能见到许多设计师搜罗来的充满趣味的物件，从废弃修车厂找来的烂铁皮，旧房子上拆下来的木门，路边淘来的石头马槽，都成了空间的一部分。把旧物与现代物器充分结合，这些别人眼中已经失去价值的东西在设计师的手中重新获得了生机，也为这个工作室增添了与众不同的趣味。

This case is the designer's studio, it is designed to create a relaxed natural working environment, to find really suitable for decorative effects designer to their needs, will be easy, casual work atmosphere into interior design.

The design of the project focus on people-oriented, designers will be their own preferences, thought, philosophy all perfusion in this more than 100 square meters of space. Here, you can see the modern style of the simple furniture, can also see the full relics of the Ming Dynasty style furniture, more can see many designers to collect interesting objects, from the abandoned garage to get rotten tin, remove the old house. The wooden door, the side of the road to the Amoy stone manger, has become a parts of space. The old and modern device fully integrated, these others have lost something of value to regain vitality in the hands of designers, but also adds the taste out of the ordinary for the studio.

神韵 新中式空间设计典藏

神韵 新中式空间设计典藏

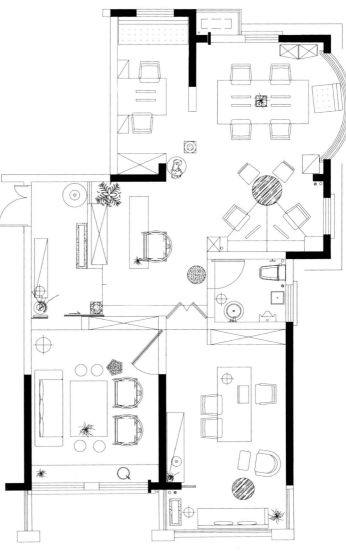

平面布置图

图书在版编目(CIP)数据

神韵：新中式空间设计典藏／海燕编．—武汉：华中科技大学出版社，2015.4
ISBN 978-7-5609-9739-1

Ⅰ．①神… Ⅱ．①海… Ⅲ．①住宅－室内装饰设计 Ⅳ．①TU241

中国版本图书馆CIP数据核字(2014)第307466号

神韵　新中式空间设计典藏

海燕　编

出版发行：华中科技大学出版社（中国·武汉）	
地　　址：武汉市武昌珞喻路1037号（邮编：430074）	
出 版 人：阮海洪	

责任编辑：赵爱华	责任监印：秦　英
责任校对：曾　晟	装帧设计：海　燕

印　　刷：深圳市雅佳图印刷有限公司
开　　本：965 mm×1270 mm　1/8
印　　张：64
字　　数：461千字
版　　次：2015年4月第1版第1次印刷
定　　价：880.00元（USD 179.99）

投稿热线：(010)64155588-8000
本书若有印装质量问题，请向出版社营销中心调换
全国免费服务热线：400-6679-118　竭诚为您服务
版权所有　侵权必究